# STUDENT'S SOLUTIONS MANUAL

## NANCY S. BOUDREAU AND JAY SCHAFFER

# ELEMENTARY STATISTICS: PICTURING THE WORLD

## SIXTH EDITION

### Ron Larson

*Pennsylvania State University*

### Betsy Farber

*Bucks County Community College*

**PEARSON**

Boston Columbus Indianapolis New York San Francisco Upper Saddle River
Amsterdam Cape Town Dubai London Madrid Milan Munich Paris Montreal Toronto
Delhi Mexico City São Paulo Sydney Hong Kong Seoul Singapore Taipei Tokyo

ISBN-13: 978-0-321-91125-4
ISBN-10: 0-321-91125-3

1 2 3 4 5 6 EBM 18 17 16 15 14

www.pearsonhighered.com

# CONTENTS

# Introduction to Statistics

## 1.1 AN OVERVIEW OF STATISTICS

### 1.1 Try It Yourself Solutions

**1a.** The population consists of the prices per gallon of regular gasoline at all gasoline stations in the United States. The sample consists of the prices per gallon of regular gasoline at the 800 surveyed stations.

  **b.** The data set consists of the 800 prices.

**2a.** Because the numerical measure of $5,150,694 is based on the entire collection of employee's salaries, it is from a population.

  **b.** Because the numerical measure is a characteristic of a population, it is a parameter.

**3a.** Descriptive statistics involve the statement "31% support their kids financially until they graduate college and 6% provide financial support until they start college."

  **b.** An inference drawn from the survey is that a higher percentage of parents support their kids financially until they graduate college.

### 1.1 EXERCISE SOLUTIONS

**1.** A sample is a subset of a population.

**3.** A parameter is a numerical description of a population characteristic. A statistic is a numerical description of a sample characteristic.

**5.** False. A statistic is a numerical measure that describes a sample characteristic.

**7.** True

**9.** False. A population is the collection of *all* outcomes, responses, measurements, or counts that are of interest.

**11.** The data set is a population because it is a collection of the revenue of each of the 30 companies in the Dow Jones Industrial Average.

**13.** The data set is a sample because the collection of the 500 spectators is a subset within the population of the stadium's 42,000 spectators.

**15.** The data set is a sample because the collection of the 20 patients is a subset of the population of 100 patients at the hospital.

**17.** The data set is a population because it is a collection of all the golfers' scores in the tournament.

**1**

**19.** The data set is a population because it is a collection of all the U.S. presidents' political parties.

**21.** Population: Parties of registered voters in Warren County

Sample: Parties of Warren County voters responding to online survey

**23.** Population: Ages of adults in the United States who own cell phones

Sample: Ages of adults in the United States who own Samsung cell phones

**25.** Population: Collection of the responses of all adults in the United States

Sample: Collection of the responses of the 1015 U.S. adults surveyed

**27.** Population: Collection of the immunization status of all adults in the U.S.

Sample: Collection of the immunization status of the 12,082 U.S. adults surveyed

**29.** Population: Collection of the average billing rates of all U.S. law firms

Sample: Collection of the average billing rates of the 55 U.S. law firms surveyed

**31.** Population: Collection of the effect of sleepiness on all pilots

Sample: Collection of the effect of sleepiness on the 202 pilots surveyed

**33.** Population: Collection of the starting salaries at all 500 companies listed in the Standard & Poor's 500

Sample: Collection of the starting salaries at the 65 companies listed in the Standard & Poor's 500 that were contacted by the researcher

**35.** Statistic. The value $68,000 is a numerical description of a sample of annual salaries.

**37.** Parameter. The 62 surviving passengers out of 97 total passengers is a numerical description of all of the passengers of the Hindenburg that survived.

**39.** Statistic. The value 8% is a numerical description of a sample of computer users.

**41.** Statistic. The value 52% is a numerical description of a sample of U.S. adults.

**43.** The statement "20% admit that they have made a serious error due to sleepiness" is an example of descriptive statistics.

An inference drawn from the sample is that an association exists between sleepiness and pilot error.

**45.** Answers will vary.

**47.** (a) An inference drawn from the sample is that senior citizens who live in Florida have better memories than senior citizens who do not live in Florida.

(b) It implies that if you live in Florida, you will have better memory.

**49.** Answers will vary.

## 1.2 DATA CLASSIFICATION

## 1.2 Try It Yourself Solutions

**1a.** One data set contains names of cities and the other contains city populations.

**b.** City names: Nonnumerical
City Populations: Numerical

**c.** City names: Qualitative
City Populations: Quantitative

**2a.** (1) The final standings represent a ranking of basketball teams.

(2) The collection of phone numbers represents labels. No mathematical computations can be made.

**b.** (1) Ordinal, because the data can be put in order.

(2) Nominal, because you cannot make calculations on the data.

**3a.** (1) The data set is the collection of body temperatures.

(2) The data set is the collection of heart rates.

**b.** (1) Interval, because the data can be ordered and meaningful differences can be calculated, but it does not make sense writing a ratio using the temperatures.

(2) Ratio, because the data can be ordered, the data can be written as a ratio, meaningful differences can be calculated, and the data set contains an inherent zero.

## 1.2 EXERCISE SOLUTIONS

**1.** Nominal and ordinal

**3.** False. Data at the ordinal level can be qualitative or quantitative.

**5.** False. More types of calculations can be performed with data at the interval level than with data at the nominal level.

**7.** Quantitative, because heights of hot air balloons are numerical measurements.

9. Qualitative, because the colors are attributes.

11. Quantitative, because weights of infants are numerical measurements.

13. Qualitative, because the poll responses are attributes.

15. Interval. Data can be ordered and meaningful differences can be calculated, but it does not make sense to say one year is a multiple of another.

17. Nominal. No mathematical computations can be made and data are categorized using numbers.

19. Ordinal. Data can be arranged in order, or ranked, but differences between data entries are not meaningful.

21. Horizontal: Ordinal; Vertical: Ratio

23. Horizontal: Nominal; Vertical: Ratio

25. (a) Interval      (b) Nominal      (c) Ratio      (d) Ordinal

27. Qualitative. Ordinal. Data can be arranged in order, but differences between data entries are not meaningful.

29. Qualitative. Nominal. No mathematical computations can be made and data are categorized using names.

31. Qualitative. Ordinal. Data can be arranged in order, but differences between data entries are not meaningful.

33. An inherent zero is a zero that implies "none." Answers will vary.

## 1.3 DATA COLLECTION AND EXPERIMENTAL DESIGN

## 1.3 Try It Yourself Solutions

1a. The study does not apply a treatment to the elk.

b. This is an observational study.

2a. There is no way to tell why people quit smoking. They could have quit smoking either from the gum or from watching the DVD. The gum and the DVD could be confounding variables.

b. Two experiments could be done; one using the gum and the other using the DVD. Or just conduct one experiment using either the gum or the DVD.

3. *Sample answers:*

a. Start with the first digits 92630782 …

**b.** 92 | 63 | 07 | 82 | 40 | 19 | 26

**c.** 63, 7, 40, 19, 26

**4a.** (1) The sample was selected by using the students in a randomly chosen class. This sampling technique is cluster sampling.

(2) The sample was selected by numbering each student in the school, randomly choosing a starting number, and selecting students at regular intervals from the starting number. This sampling technique is systematic sampling.

**b.** (1) The sample may be biased because some classes may be more familiar with stem cell research than other classes and have stronger opinions.

(2) The sample may be biased if there is any regularly occurring pattern in the data.

## 1.3 EXERCISE SOLUTIONS

1. In an experiment, a treatment is applied to part of a population and responses are observed. In an observational study, a researcher measures characteristics of interest of a part of a population but does not change existing conditions.

3. In a random sample, every member of the population has an equal chance of being selected. In a simple random sample, every possible sample of the same size has an equal chance of being selected.

5. False. A placebo is a fake treatment.

7. False. Using stratified sampling guarantees that members of each group within a population will be sampled.

9. False. To select a systematic sample, a population is ordered in some way and then members of the population are selected at regular intervals.

11. Observational study. The study does not attempt to influence the responses of the subjects and there is no treatment.

13. Experiment. The study applies a treatment (different genres of music) to the subjects.

15. (a) The experimental units are the 250 females ages 30–35 in the study. The treatment is the new allergy drug.

(b) A problem with the design is that there may be some bias on the part of the researcher if the researcher knows which patients were given the real drug. A way to eliminate this problem would be to make the study into a double-blind experiment.

(c) The study would be a double-blind study if the researcher did not know which patients received the real drug or the placebo.

**17.** Answers will vary. *Sample answer*: Starting at the left-most number in row 6:
28/70/35/17/09/94/45/64/83/96/73/78/
The numbers would be 28,70,35,17,9,94,45,64,83,96,73,78.

**19.** Answers will vary.

**21.** Answers will vary. *Sample answer*: Number the volunteers from 1 to 18. Using the random number table in Appendix B, starting with the left-most number in row 16:
29/55/31/84/32/**13**/63/00/55/29/**02**/79/**18**/**10**/**17**/49/02/77/90/31/50/91/20/93/99
23/50/**12**/26/42/63/**08**/10/81/91/89/42/**06**/78/00/55/13/75/47/**07**/
Treatment group: Maria, Adam, Bridget, Carlos, Susan, Rick, Dan, Mary, and Connie.
Control group: Jake, Mike, Lucy, Ron, Steve, Vanessa, Kate, Pete, and Judy.

**23.** Simple random sampling is used because each telephone number has an equal chance of being dialed, and all samples of 1400 phone numbers have an equal chance of being selected. The sample may be biased because telephone sampling only samples those individuals who have telephones, who are available, and who are willing to respond.

**25.** Convenience sampling is used because the students are chosen due to their convenience of location. Bias may enter into the sample because the students sampled may not be representative of the population of students. For example, there may be an association between time spent at the library and drinking habits.

**27.** Simple random sampling is used because each customer has an equal chance of being contacted, and all samples of 580 customers have an equal chance of being selected.

**29.** Stratified sampling is used because a sample is taken from each one-acre subplot (stratum).

**31.** Census, because it is relatively easy to obtain the ages of the 115 residents

**33.** The question is biased because it already suggests that eating whole-grain foods improves your health. The question might be rewritten as "How does eating whole-grain foods affect your health?"

**35.** The survey question is unbiased because it does not imply how much exercise is good or bad.

**37.** The households sampled represent various locations, ethnic groups, and income brackets. Each of these variables is considered a stratum. Stratified sampling ensures that each segment of the population is represented.

**39.** Open Question
Advantage: Allows respondent to express some depth and shades of meaning in the answer. Allows for new solutions to be introduced.
Disadvantage: Not easily quantified and difficult to compare surveys.

Closed Question
Advantage: Easy to analyze results.
Disadvantage: May not provide appropriate alternatives and may influence the opinion of the respondent.

**41.** Answers will vary.

## CHAPTER 1 REVIEW EXERCISE SOLUTIONS

1.  Population: Collection of the responses of all U.S. adults

    Sample: Collection of the responses of the 1503 U.S. adults that were sampled

3.  Population: Collection of the responses of all U.S. adults

    Sample: Collection of the responses of the 2311 U.S. adults that were sampled

5.  Parameter. The value $2,940,657,192 is a numerical description of the total player salary for all players in Major League Baseball.

7.  Parameter. The 10 students minoring in physics is a numerical description of all math majors at a university.

9.  The statement "84% have seen a health care provider at least once in the past year" is an example of descriptive statistics. An inference drawn from the sample is that most people have gone to a health care provider at least once in the past year.

11. Quantitative, because ages are numerical measurements.

13. Quantitative, because revenues are numerical measures.

15. Interval. The data can be ordered and meaningful differences can be calculated, but it does not make sense to say that 87 degrees is 1.16 times as hot as 75 degrees.

17. Nominal. The data are qualitative and cannot be arranged in a meaningful order.

19. Experiment. The study applies a treatment (hypothyroidism drug) to the subjects.

21. *Sample answer:* The subjects could be split into male and female and then be randomly assigned to each of the five treatment groups.

23. Simple random sampling is used because random telephone numbers were generated and called. A potential source of bias is that telephone sampling only samples individuals who have telephones, who are available, and who are willing to respond.

25. Cluster sampling is used because each community is considered a cluster and every pregnant woman in a selected community is surveyed.

27. Stratified sampling is used because the population is divided by grade level and then 25 students are randomly selected from each grade level.

29. Answers will vary. *Sample answer*: Using the random number table in Appendix B, starting with the left-most number in row 5:
    063/487/693/890/379/513/925/588/771/015/092/097/
    The random numbers are 63,487,379,513,588,15,92,97.

## CHAPTER 1 QUIZ SOLUTIONS

1.  Population: Collection of the prostate conditions of all men

    Sample: Collection of the prostate conditions of the 20,000 men in the study

2.  (a) Statistic. The value 40% is a numerical description of a sample of U.S. adults.

    (b) Parameter. The 90% of members that approved the contract of the new president is a numerical description of all Board of Trustees members.

    (c) Statistic. The value 17% is a numerical description of a sample of small business owners.

3.  (a) Qualitative, because debit card pin numbers are labels and it does not make sense to find differences between numbers.

    (b) Quantitative, because final scores is a numerical measurements.

4.  (a) Ordinal, because badge numbers can be ordered and often indicate seniority of service, but no meaningful mathematical computation can be performed.

    (b) Ratio, because horsepower of one car can be expressed as a multiple of another.

    (c) Ordinal, because data can be arranged in order, but the differences between data entries make no sense.

    (d) Interval, because meaningful differences between years can be calculated, but a zero entry is not an inherent zero.

5.  (a) Observational study. The study does not attempt to influence the responses of the subjects and there is no treatment.

    (b) Experiment. The study applies a treatment (multivitamin) to the subjects.

6.  Randomized block design

7.  (a) Convenience sampling is used because all the people sampled are in one convenient location.

    (b) Systematic sampling is used because every tenth part is sampled.

    (c) Stratified sample is used because the population is first stratified and then a sample is collected from each stratum.

8.  Convenience sampling. People at campgrounds may be strongly against air pollution because they are at an outdoor location.

# Descriptive Statistics

## 2.1 Try It Yourself Solutions

**1a.** The number of classes is 7.

**b.** Min = 26, Max = 86, Class width $= \dfrac{\text{Range}}{\text{Number of classes}} = \dfrac{86 - 26}{7} = 8.57 \Rightarrow 9$

**c.**

| Lower limit | Upper limit |
|---|---|
| 26 | 34 |
| 35 | 43 |
| 44 | 52 |
| 53 | 61 |
| 62 | 70 |
| 71 | 79 |
| 80 | 88 |

**de.**

| Class | Frequency, $f$ |
|---|---|
| 26-34 | 2 |
| 35-43 | 5 |
| 44-52 | 12 |
| 53-61 | 18 |
| 62-70 | 11 |
| 71-79 | 1 |
| 80-88 | 1 |

**2a.** See part (b).

**b.**

| Class | Frequency, $f$ | Midpoint | Relative frequency | Cumulative frequency |
|---|---|---|---|---|
| 26-34 | 2 | 30 | 0.04 | 2 |
| 35-43 | 5 | 39 | 0.10 | 7 |
| 44-52 | 12 | 48 | 0.24 | 19 |
| 53-61 | 18 | 57 | 0.36 | 37 |
| 62-70 | 11 | 66 | 0.22 | 48 |
| 71-79 | 1 | 75 | 0.02 | 49 |
| 80-88 | 1 | 84 | 0.02 | 50 |
| | $\sum f = 50$ | | $\sum \dfrac{f}{n} = 1$ | |

**c.** *Sample answer*: The most common age bracket for the 50 most powerful women is 53-61. Eighty-six percent of the 50 most powerful women are older than 43. Four percent of the 50 most powerful women are younger than 35.

**9**

**3a.**

| Class Boundaries |
|:---:|
| 25.5-34.5 |
| 34.5-43.5 |
| 43.5-52.5 |
| 52.5-61.5 |
| 61.5-70.5 |
| 70.5-79.5 |
| 79.5-88.5 |

**b.** Use class midpoints for the horizontal scale and frequency for the vertical scale. (Class boundaries can also be used for the horizontal scale.)

**c.**

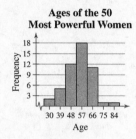

**d.** Same as 2(c).

**4a.** Same as 3(b).

**bc.**

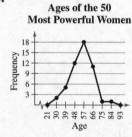

**d.** The frequency of ages increases up to 57 years old and then decreases.

**5abc.**

**6a.** Use upper class boundaries for the horizontal scale and cumulative frequency for the vertical scale.

**bc.**

**Ages of the 50 Most Powerful Women**

*Sample answer*: The greatest increase in cumulative frequency occurs between 52.5 and 61.5

**7a.** Enter data.

**b.**

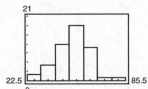

## 2.1 EXERCISE SOLUTIONS

1. Organizing the data into a frequency distribution may make patterns within the data more evident. Sometimes it is easier to identify patterns of a data set by looking at a graph of the frequency distribution.

3. Class limits determine which numbers can belong to that class.
Class boundaries are the numbers that separate classes without forming gaps between them.

5. The sum of the relative frequencies must be 1 or 100% because it is the sum of all portions or percentages of the data.

7. False. Class width is the difference between the lower (or upper limits) of consecutive classes.

9. False. An ogive is a graph that displays cumulative frequencies.

11. Class width $= \dfrac{\text{Range}}{\text{Number of classes}} = \dfrac{64-9}{7} \approx 7.9 \Rightarrow 8$
Lower class limits: 9, 17, 25, 33, 41, 49, 57
Upper class limits: 16, 24, 32, 40, 48, 56, 64

13. Class width $= \dfrac{\text{Range}}{\text{Number of classes}} = \dfrac{135-17}{8} = 14.75 \Rightarrow 15$
Lower class limits: 17, 32, 47, 62, 77, 92, 107, 122
Upper class limits: 31, 46, 61, 76, 91, 106, 121, 136

**15.** (a) Class width $= 31 - 20 = 11$

(b) and (c)

| Class | Midpoint | Class boundaries |
|-------|----------|------------------|
| 20-30 | 25 | 19.5-30.5 |
| 31-41 | 36 | 30.5-41.5 |
| 42-52 | 47 | 41.5-52.5 |
| 53-63 | 58 | 52.5-63.5 |
| 64-74 | 69 | 63.5-74.5 |
| 75-85 | 80 | 74.5-85.5 |
| 86-96 | 91 | 85.5-96.5 |

**17.**

| Class | Frequency, $f$ | Midpoint | Relative frequency | Cumulative frequency |
|-------|----------------|----------|--------------------|----------------------|
| 20-30 | 19 | 25 | 0.05 | 19 |
| 31-41 | 43 | 36 | 0.12 | 62 |
| 42-52 | 68 | 47 | 0.19 | 130 |
| 53-63 | 69 | 58 | 0.19 | 199 |
| 64-74 | 74 | 69 | 0.20 | 273 |
| 75-85 | 68 | 80 | 0.19 | 341 |
| 86-96 | 24 | 91 | 0.07 | 365 |
| | $\sum f = 365$ | | $\sum \dfrac{f}{n} \approx 1$ | |

**19.** (a) Number of classes = 7    (b) Least frequency $\approx 10$
(c) Greatest frequency $\approx 300$    (d) Class width = 10

**21.** (a) 50        (b) 345.5-365.5 pounds

**23.** (a) 15        (b) 385.5 pounds
(c) $31 - 6 = 25$    (d) $50 - 42 = 8$

**25.** (a) Class with greatest relative frequency: 39-40 centimeters
Class with least relative frequency: 34-35 centimeters

(b) Greatest relative frequency $\approx 0.25$
Least relative frequency $\approx 0.02$

(c) Approximately 0.08

**27.** Class with greatest frequency: 29.5-32.5
Classes with least frequency: 11.5-14.5 and 38.5-41.5

**29.** Class width $= \dfrac{\text{Range}}{\text{Number of classes}} = \dfrac{39-0}{5} = 7.8 \Rightarrow 8$

| Class | Frequency, $f$ | Midpoint | Relative frequency | Cumulative frequency |
|---|---|---|---|---|
| 0-7 | 8 | 3.5 | 0.32 | 8 |
| 8-15 | 8 | 11.5 | 0.32 | 16 |
| 16-23 | 3 | 19.5 | 0.12 | 19 |
| 24-31 | 3 | 27.5 | 0.12 | 22 |
| 32-39 | 3 | 35.5 | 0.12 | 25 |
| | $\sum f = 25$ | | $\sum \dfrac{f}{n} = 1$ | |

Classes with greatest frequency: 0-7, 8-15
Classes with least frequency: 16-23, 24-31, 32-39

**31.** Class width $= \dfrac{\text{Range}}{\text{Number of classes}} = \dfrac{7119-1000}{6} \approx 1019.8 \Rightarrow 1020$

| Class | Frequency, $f$ | Midpoint | Relative frequency | Cumulative frequency |
|---|---|---|---|---|
| 1000-2019 | 12 | 1509.5 | 0.5455 | 12 |
| 2020-3039 | 3 | 2529.5 | 0.1364 | 15 |
| 3040-4059 | 2 | 3549.5 | 0.0909 | 17 |
| 4060-5079 | 3 | 4569.5 | 0.1364 | 20 |
| 5080-6099 | 1 | 5589.5 | 0.0455 | 21 |
| 6100-7119 | 1 | 6609.5 | 0.0455 | 22 |
| | $\sum f = 22$ | | $\sum \dfrac{f}{n} \approx 1$ | |

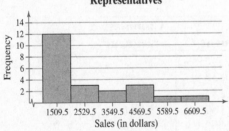

**July Sales for Representatives**

*Sample answer*: The graph shows that most of the sales representatives at the company sold between \$1000 and \$2019.

**33.** Class width $= \dfrac{\text{Range}}{\text{Number of classes}} = \dfrac{514 - 291}{8} = 27.875 \Rightarrow 28$

| Class | Frequency, $f$ | Midpoint | Relative frequency | Cumulative frequency |
|---|---|---|---|---|
| 291-318 | 5 | 304.5 | 0.1667 | 5 |
| 319-346 | 4 | 332.5 | 0.1333 | 9 |
| 347-374 | 3 | 360.5 | 0.1000 | 12 |
| 375-402 | 5 | 388.5 | 0.1667 | 17 |
| 403-430 | 6 | 416.5 | 0.2000 | 23 |
| 431-458 | 4 | 444.5 | 0.1333 | 27 |
| 459-486 | 1 | 472.5 | 0.0333 | 28 |
| 487-514 | 2 | 500.5 | 0.0667 | 30 |
| | $\sum f = 30$ | | $\sum \dfrac{f}{n} = 1$ | |

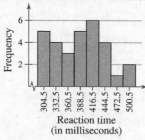

**Reaction Times for Females**

*Sample answer*: The graph shows that the most frequent reaction times were between 403 and 430 milliseconds.

**35.** Class width $= \dfrac{\text{Range}}{\text{Number of classes}} = \dfrac{10 - 1}{5} = 1.8 \Rightarrow 2$

| Class | Frequency, $f$ | Midpoint | Relative frequency | Cumulative frequency |
|---|---|---|---|---|
| 1-2 | 2 | 1.5 | 0.0833 | 2 |
| 3-4 | 2 | 3.5 | 0.0833 | 4 |
| 5-6 | 5 | 5.5 | 0.2083 | 9 |
| 7-8 | 10 | 7.5 | 0.4167 | 19 |
| 9-10 | 5 | 9.5 | 0.2083 | 24 |
| | $\sum f = 24$ | | $\sum \dfrac{f}{n} \approx 1$ | |

**Taste Test Ratings**

Class with greatest relative frequency: 7-8
Class with least relative frequency: 1-2 and 3-4

**37.** Class width $= \dfrac{\text{Range}}{\text{Number of classes}} = \dfrac{547 - 417}{5} = 26 \Rightarrow 27$

| Class | Frequency, $f$ | Midpoint | Relative frequency | Cumulative frequency |
|---|---|---|---|---|
| 417-443 | 5 | 430 | 0.20 | 5 |
| 444-470 | 5 | 457 | 0.20 | 10 |
| 471-497 | 6 | 484 | 0.24 | 16 |
| 498-524 | 4 | 511 | 0.16 | 20 |
| 525-551 | 5 | 538 | 0.20 | 25 |
| | $\sum f = 25$ | | $\sum \dfrac{f}{n} = 1$ | |

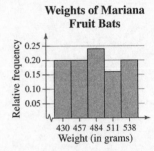

**Weights of Mariana Fruit Bats**

Class with greatest relative frequency: 471-497
Class with least relative frequency: 498-524

**39.** Class width $= \dfrac{\text{Range}}{\text{Number of classes}} = \dfrac{73 - 52}{6} = 3.5 \Rightarrow 4$

| Class | Frequency, $f$ | Relative frequency | Cumulative frequency |
|---|---|---|---|
| 52-55 | 3 | 0.125 | 3 |
| 56-59 | 3 | 0.125 | 6 |
| 60-63 | 9 | 0.375 | 15 |
| 64-67 | 4 | 0.167 | 19 |
| 68-71 | 4 | 0.167 | 23 |
| 72-75 | 1 | 0.042 | 24 |
| | $\sum f = 24$ | $\sum \dfrac{f}{n} \approx 1$ | |

**Retirement Ages**

Location of the greatest increase in frequency: 60-63

**41.** Class width $= \dfrac{\text{Range}}{\text{Number of classes}} = \dfrac{15-0}{6} = 2.5 \Rightarrow 3$

| Class | Frequency, $f$ | Midpoint | Relative frequency | Cumulative frequency |
|---|---|---|---|---|
| 0-2 | 17 | 1 | 0.3953 | 17 |
| 3-5 | 17 | 4 | 0.3953 | 34 |
| 6-8 | 7 | 7 | 0.1628 | 41 |
| 9-11 | 1 | 10 | 0.0233 | 42 |
| 12-14 | 0 | 13 | 0.0000 | 42 |
| 15-17 | 1 | 16 | 0.0233 | 43 |
| | $\sum f = 43$ | | $\sum \dfrac{f}{N} = 1$ | |

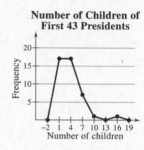

**Number of Children of First 43 Presidents**

*Sample answer:* The graph shows that most of the first 43 presidents had fewer than 6 children.

**43. (a)** Class width $= \dfrac{\text{Range}}{\text{Number of classes}} = \dfrac{120-65}{6} \approx 9.2 \Rightarrow 10$

| Class | Frequency, $f$ | Midpoint | Relative frequency | Cumulative frequency |
|---|---|---|---|---|
| 65-74 | 4 | 69.5 | 0.1667 | 4 |
| 75-84 | 7 | 79.5 | 0.2917 | 11 |
| 85-94 | 4 | 89.5 | 0.1667 | 15 |
| 95-104 | 5 | 99.5 | 0.2083 | 20 |
| 105-114 | 3 | 109.5 | 0.1250 | 23 |
| 115-124 | 1 | 119.5 | 0.0417 | 24 |
| | $\sum f = 24$ | | $\sum \dfrac{f}{N} \approx 1$ | |

**(b)**

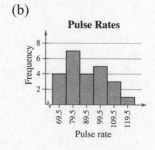

**Pulse Rates**

**(c)**

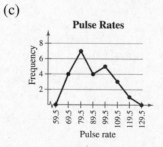

**Pulse Rates**

(d)

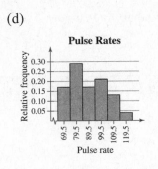

**Pulse Rates**

(e)

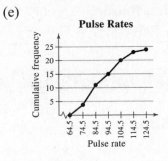

**Pulse Rates**

**45.** (a) $\text{Class width} = \dfrac{\text{Range}}{\text{Number of classes}} = \dfrac{104 - 61}{8} = 5.375 \Rightarrow 6$

| Class | Frequency, $f$ | Midpoint | Relative frequency |
|---|---|---|---|
| 61-66 | 1 | 63.5 | 0.033 |
| 67-72 | 3 | 69.5 | 0.100 |
| 73-78 | 6 | 75.5 | 0.200 |
| 79-84 | 10 | 81.5 | 0.333 |
| 85-90 | 5 | 87.5 | 0.167 |
| 91-96 | 2 | 93.5 | 0.067 |
| 97-102 | 2 | 99.5 | 0.067 |
| 103-108 | 1 | 105.5 | 0.033 |
| | $\sum f = 30$ | | $\sum \dfrac{f}{N} = 1$ |

**Daily Withdrawals**

(b) 16.7%, because the sum of the relative frequencies for the last three classes is 0.167.

(c) $9700, because the sum of the relative frequencies for the last two classes is 0.10.

**47.**

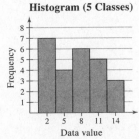

**Histogram (5 Classes)**

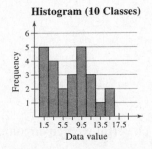

**Histogram (10 Classes)**

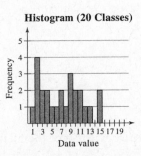

**Histogram (20 Classes)**

In general, a greater number of classes better preserves the actual values of the data set but is not as helpful for observing general trends and making conclusions. In choosing the number of classes, an important consideration is the size of the data set. For instance, you would not want to use 20 classes if your data set contained 20 entries. In this particular example, as the number of classes increases, the histogram shows more fluctuation. The histograms with 10 and 20 classes have classes with zero frequencies. Not much is gained by using more than five classes. Therefore, it appears that five classes would be best.

## 2.2 MORE GRAPHS AND DISPLAYS

### 2.2 Try It Yourself Solutions

**1a.**
```
2|
3|
4|
5|
6|
7|
8|
```
**b.**
```
2|6
3|1  5  7
4|3  3  3  4  5  7  8  8  9
5|0  1  1  1  1  2  4  4  4  4  5  5  5  6  7  7  7  8  8  8  8  9  9  9
6|2  2  3  4  5  5  5  6  6  7  7
7|2
8|6
```

Key 3|6 = 36

**c.** *Sample answer:* Most of the most powerful women are between 40 and 70 years old.

**2a, b.**
```
2|
2|6
3|1
3|5  7
4|3  3  3  4
4|5  7  8  8  9
5|0  1  1  1  1  2  4  4  4  4
5|5  5  5  6  7  7  7  8  8  8  8  9  9  9
6|2  2  3  4
6|5  5  5  6  6  7  7
7|2
7|
8|
8|6
```

Key  3|5 = 35

**c.** *Sample answer:*  Most of the 50 most powerful women are older than 50.

**3a.** Use the age for the horizontal axis.
  **b.**

Ages of the 50 Most Powerful Women

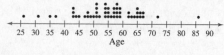

c. *Sample answer:* Most of the ages cluster between 43 and 67 years old. The age of 86 years old is an unusual data entry.

**4a.**

| Type of Degree | $f$ | Relative Frequency | Angle |
|---|---|---|---|
| Associate's | 455 | 0.235 | 85° |
| Bachelor's | 1051 | 0.542 | 195° |
| Master's | 330 | 0.170 | 61° |
| Doctoral | 104 | 0.054 | 19° |
| | $\sum f = 1940$ | $\sum \dfrac{f}{N} \approx 1$ | $\sum = 360°$ |

**b.**

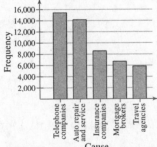

Earned Degrees Conferred in 1990

Associate's 23.5%
Doctoral 5.4%
Master's 17.0%
Bachelor's 54.2%

c. From 1990 to 2011, as percentages of total degrees conferred, associate's degrees increased by 3%, bachelor's degrees decreased by 5.9%, master's degrees increased by 3.6%, and doctoral degrees decreased by 0.8%.

**5a.**

| Cause | Frequency, $f$ |
|---|---|
| Auto repair and service | 14,156 |
| Insurance companies | 8,568 |
| Mortgage brokers | 6,712 |
| Telephone companies | 15,394 |
| Travel agencies | 5,841 |

**b.**

Causes of BBB Complaints

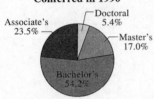

c. *Sample answer:* Telephone companies and auto repair and service account for over half of all complaints received by the BBB.

**6a, b.**

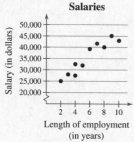

**c.** It appears that the longer an employee is with the company, the larger the employee's salary will be.

**7ab.**

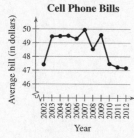

**c.** The average bill increased from 2002 to 2003, then it hovered from 2003 to 2009, and decreased from 2009 to 2012.

## 2.2 EXERCISE SOLUTIONS

**1.** Quantitative: stem-and-leaf plot, dot plot, histogram, time series chart, scatter plot.
Qualitative: pie chart, Pareto chart

**3.** Both the stem-and-leaf plot and the dot plot allow you to see how data are distributed, determine specific data entries, and identify unusual data values.

**5. b**         **7. a**

**9.** 27, 32, 41, 43, 43, 44, 47, 47, 48, 50, 51, 51, 52, 53, 53, 53, 54, 54, 54, 54, 55, 56, 56, 58, 59, 68, 68, 68, 73, 78, 78, 85
Max: 85    Min: 27

**11.** 13, 13, 14, 14, 14, 15, 15, 15, 15, 15, 16, 17, 17, 18, 19
Max: 19    Min: 13

**13.** *Sample answer*: Users spend the most amount of time on Facebook and the least amount of time on LinkedIn.

**15.** *Sample answer*: Tailgaters irk drivers the most, while too-cautious drivers irk drivers the least.

**17. Exam Scores**     Key: 6|7 = 67

```
6 | 7  8
7 | 3  5  5  6  9
8 | 0  0  2  3  5  5  7  7  8
9 | 0  1  1  1  2  4  5  5
```

*Sample answer*: Most grades for the biology midterm were in the 80s and 90s.

**19. Ice Thickness (in centimeters)**   Key: 4|3 = 4.3

```
4 | 3  9
5 | 1  8  8  8  9
6 | 4  8  9  9  9
7 | 0  0  2  2  2  5
8 | 0  1
```

*Sample answer:*  Most of the ice had a thickness of 5.8 centimeters to 7.2 centimeters.

**21. Ages of Highest-Paid CEOs**     Key: 5|0 = 50

```
5 | 0  2  3
5 | 5  5  6  6  7  7  8  8  8  8  9  9  9  9  9
6 | 0  1  1  1  3  4
6 | 5  5  6  6  7
7 | 2
```

*Sample answer:*  Most of the highest-paid CEOs have ages that range from 55 and 64 years old.

**23.**

**Systolic Blood Pressures**

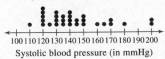

100 110 120 130 140 150 160 170 180 190 200
Systolic blood pressure (in mmHg)

*Sample answer:*  Systolic blood pressure tends to be between 120 and 150 millimeters of mercury.

**25.**   **How Will You Invest in 2013?**

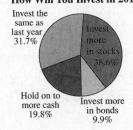

Invest the same as last year 31.7%

Invest more in stocks 38.6%

Hold on to more cash 19.8%

Invest more in bonds 9.9%

*Sample answer:*  The majority of people will either invest more in stocks or invest the same as last year.

**27.**

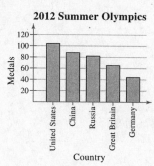

*Sample answer:* The United States won the most medals out of the five countries and Germany won the least.

**29.**

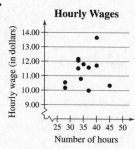

*Sample answer:* It appears that there is no relation between wages and hours worked.

**31.**

*Sample answer:* The number of motorcycle registrations has increased from 2000 to 2011.

**33.**

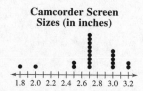

The stem-and-leaf plot helps you see that most values are from 2.5 to 3.2. The dot plot helps you see that the values 2.7 and 3.0 occur most frequently, with 2.7 occurring most frequently.

**35.**

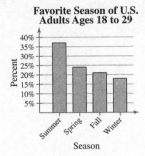

Favorite Season of U.S.
Adults Ages 18 to 29

The pie chart helps you to see the percentages as parts of a whole, with summer being the largest. It is also shows that while summer is the largest percentage, it only makes up about one-third of the pie chart. That means that about two-thirds of U.S. adults ages 18 to 29 prefer a season other than summer. This means it would not be a fair statement to say that most U.S. adults ages 18 to 29 prefer summer. The Pareto chart helps you to see the rankings of the seasons. It helps you to see that the favorite seasons in order from greatest to least percentage are summer, spring, fall, and winter.

**37.** (a) The graph is misleading because the large gap from 0 to 90 makes it appear that the sales for the 3rd quarter are disproportionately larger than the other quarters.

  (b)

Sales for Company A

**39.** (a) The graph is misleading because the angle makes it appear as though the 3rd quarter had a larger percent of sales than the others, when the 1st and 3rd quarters have the same percent.

  (b)

Sales for Company B

**41.** (a) At Law Firm A, the lowest salary was $90,000 and the highest salary was $203,000. At Law Firm B, the lowest salary was $90,000 and the highest salary was $190,000.

  (b) There are 30 lawyers at Law Firm A and 32 lawyers at Law Firm B.

  (c) At Law Firm A, the salaries tend to be clustered at the far ends of the distribution range. At Law Firm B, the salaries are spread out.

## 2.3 MEASURES OF CENTRAL TENDENCY

### 2.3 Try It Yourself Solutions

**1a.** $\sum x = 74 + 78 + 81 + 87 + 81 + 80 + 77 + 80 + 85 + 78 + 80 + 83 + 75 + 81 + 73 = 1193$

**b.** $\bar{x} = \dfrac{\sum x}{n} = \dfrac{1193}{15} \approx 79.5$

**c.** The mean height of the player is about 79.5 inches.

**2a.** 18, 18, 19, 19, 19, 20, 21, 21, 21,21, 23, 24, 24, 26, 27, 27, 29, 30, 30, 30, 33, 33, 34, 35, 38

**b.** median = 24

**c.** The median age of the sample of fans at the concert is 24.

**3a.** 10, 50, 50, 70, 70, 80, 100, 100, 120, 130

**b.** median $= \dfrac{70 + 80}{2} = \dfrac{150}{2} = 75$

**c.** The median price of the sample of digital photo frames is $75.

**4a.** 324, 385, 450, 450, 462, 475, 540, 540, 564, 618, 624, 638, 670, 670, 670, 705, 720, 723, 750, 750, 825, 830, 912, 975, 980, 980, 1100, 1260, 1420, 1650

**b.** The price that occurs with the greatest frequency is $670 per square foot.

**c.** The mode of the prices for the sample of South Beach, FL condominiums is $670 per square foot.

**5a.** "Better prices" occurs with the greatest frequency (399).

**b.** In this sample, there were more people who shop online for better prices than for any other reason.

**6a.** $\bar{x} = \dfrac{\sum x}{n} = \dfrac{410}{19} \approx 21.6$

median = 21
mode = 20

**b.** The mean in Example 6 ($\bar{x} \approx 23.8$) was heavily influenced by the entry 65. Neither the median nor the mode was affected as much by the entry 65.

**7a, b.**

| Source | Score, $x$ | Weight, $w$ | $x \cdot w$ |
|---|---|---|---|
| Test mean | 86 | 0.50 | 43.0 |
| Midterm | 96 | 0.15 | 14.4 |
| Final exam | 98 | 0.20 | 19.6 |
| Computer lab | 98 | 0.10 | 9.8 |
| Homework | 100 | 0.05 | 5.0 |
| | | $\sum w = 1$ | $\sum(x \cdot w) = 91.8$ |

**c.** $\bar{x} = \dfrac{\sum(x \cdot w)}{\sum w} = \dfrac{91.8}{1} = 91.8$

**d.** The weighted mean for the course is 91.8. So, you did get an A.

**8a, b, c.**

| Class | Midpoint, $x$ | Frequency, $f$ | $x \cdot f$ |
|-------|---------------|----------------|-------------|
| 26-34 | 30 | 2 | 60 |
| 35-43 | 39 | 5 | 195 |
| 44-52 | 48 | 12 | 576 |
| 53-61 | 57 | 18 | 1026 |
| 62-70 | 66 | 11 | 726 |
| 71-79 | 75 | 1 | 75 |
| 80-88 | 84 | 1 | 84 |
| | | $N = 50$ | $\sum(x \cdot f) = 2742$ |

**d.** $\mu = \dfrac{\sum(x \cdot f)}{N} = \dfrac{2742}{50} \approx 54.8$

## 2.3 EXERCISE SOLUTIONS

**1.** True

**3.** True

**5.** *Sample answer:* 1, 2, 2, 2, 3

**7.** *Sample answer:* 2, 5, 7, 9, 35

**9.** The shape of the distribution is skewed right because the bars have a "tail" to the right.

**11.** The shape of the distribution is uniform because the bars are approximately the same height.

**13.** (11), because the distribution values range from 1 to 12 and has (approximately) equal frequencies.

**15.** (12), because the distribution has a maximum value of 90 and is skewed left due to a few students scoring much lower than the majority of the students.

**17.** $\bar{x} = \dfrac{\sum x}{n} = \dfrac{192}{13} \approx 14.8$

12  12  13  14  14  15  **15**  15  16  16  16  16  18
median = 15

mode = 16 (occurs 4 times)

**19.** $\bar{x} = \dfrac{\sum x}{n} = \dfrac{8249}{7} \approx 1178.4$

818  1125  1155  **1229**  1275  1277  1370
median = 1229

mode = none
The mode cannot be found because no data entry is repeated.

**21.** $\bar{x} = \dfrac{\sum x}{n} = \dfrac{603}{14} \approx 43.1$

39 40 41 42 42 42 44 44 44 44 44 45 45 47

$$\text{median} = \frac{44 + 44}{2} = 44$$

mode = 44 (occurs 5 times)

**23.** $\bar{x} = \dfrac{\sum x}{n} = \dfrac{481}{16} \approx 30.1$

17 21 21 25 26 30 31 31 31 34 34 34 35 36 37 38

$$\text{median} = \frac{31 + 31}{2} = 31$$

mode = 31 and 34 (both occur 3 times)

**25.** $\bar{x} = \dfrac{\sum x}{n} = \dfrac{83}{5} = 16.6$

1.0   10.0   **15.0**   25.5   31.5

↖ median = 43

mode = none
The mode cannot be found because no data entry is repeated.

**27.** $\bar{x}$ is not possible (nominal data)
median = not possible (nominal data)
mode = "Eyeglasses"
The mean and median cannot be found because the data are at the nominal level of measurement.

**29.** $\bar{x}$ is not possible (nominal data)
median is not possible (nominal data)
mode = "Junior"
The mean and median cannot be found because the data are at the nominal level of measurement.

**31.** $\bar{x} = \dfrac{\sum x}{n} = \dfrac{835}{28} \approx 29.8$

6 7 12 15 18 19 20 24 24 24 25 28 29 32 32 33 35 35 36 38 39 40 41 42 47 48 51

$$\text{median} = \frac{32 + 32}{2} = 32$$

mode = 24, 35 (both occur 3 times each)

**33.** $\bar{x} = \dfrac{\sum x}{n} = \dfrac{292}{15} \approx 19.5$

5  8  10  15  15  15  17  **20**  21  22  22  25  28  32  37

↖ median = 20

mode = 15 (occurs 3 times)

**35.** The data are skewed right.

A = mode, because it is the data entry that occurred most often.

B = median, because the median is to the left of the mean in a skewed right distribution.

C = mean, because the mean is to the right of the median in a skewed right distribution.

**37.** Mode, because the data are at the nominal level of measurement.

**39.** Mean, because the distribution is symmetric and there are no outliers.

**41.**

| Source | Score, $x$ | Weight, $w$ | $x \cdot w$ |
|---|---|---|---|
| Homework | 85 | 0.05 | 4.25 |
| Quiz | 80 | 0.35 | 28 |
| Project | 100 | 0.20 | 20 |
| Speech | 90 | 0.15 | 13.5 |
| Final exam | 93 | 0.25 | 23.25 |
| | | $\sum w = 1$ | $\sum(x \cdot w) = 89$ |

$$\bar{x} = \frac{\sum(x \cdot w)}{\sum w} = \frac{89}{1} = 89$$

**43.**

| Balance, $x$ | Days, $w$ | $x \cdot w$ |
|---|---|---|
| $523 | 24 | 12,552 |
| $2415 | 2 | 4830 |
| $250 | 4 | 1000 |
| | $\sum w = 30$ | $\sum(x \cdot w) = 18,382$ |

$$\bar{x} = \frac{\sum(x \cdot w)}{\sum w} = \frac{18,382}{30} \approx \$612.73$$

**45.**

| Grade | Points, $x$ | Credits, $w$ | $x \cdot w$ |
|---|---|---|---|
| A | 4 | 4 | 16 |
| B | 3 | 3 | 9 |
| B | 3 | 3 | 9 |
| C | 2 | 3 | 6 |
| D | 1 | 2 | 2 |
| | | $\sum w = 15$ | $\sum(x \cdot w) = 42$ |

$$\bar{x} = \frac{\sum(x \cdot w)}{\sum w} = \frac{42}{15} = 2.8$$

**47.**

| Source | Score, $x$ | Weight, $w$ | $x \cdot w$ |
|--------|-----------|-------------|-------------|
| Homework | 85 | 0.05 | 4.25 |
| Quiz | 80 | 0.35 | 28 |
| Project | 100 | 0.20 | 20 |
| Speech | 90 | 0.15 | 13.5 |
| Final exam | 85 | 0.25 | 21.25 |
| | | $\sum w = 1$ | $\sum (x \cdot w) = 87$ |

$$\bar{x} = \frac{\sum (x \cdot w)}{\sum w} = \frac{87}{1} = 87$$

**49.**

| Class | Midpoint, $x$ | Frequency, $f$ | $x \cdot f$ |
|-------|---------------|----------------|-------------|
| 29-33 | 31 | 11 | 341 |
| 34-38 | 36 | 12 | 432 |
| 39-43 | 41 | 2 | 82 |
| 44-48 | 46 | 5 | 230 |
| | | $n = 30$ | $\sum (x \cdot f) = 1085$ |

$$\bar{x} = \frac{\sum (x \cdot f)}{n} = \frac{1085}{30} \approx 36.2 \text{ miles per gallon}$$

**51.**

| Class | Midpoint, $x$ | Frequency, $f$ | $x \cdot f$ |
|-------|---------------|----------------|-------------|
| 0-9 | 4.5 | 44 | 198.0 |
| 10-19 | 14.5 | 66 | 957.0 |
| 20-29 | 24.5 | 32 | 784.0 |
| 30-39 | 34.5 | 53 | 1828.5 |
| 40-49 | 44.5 | 35 | 1557.5 |
| 50-59 | 54.5 | 31 | 1689.5 |
| 60-69 | 64.5 | 23 | 1483.5 |
| 70-79 | 74.5 | 13 | 968.5 |
| 80-89 | 84.5 | 2 | 169.0 |
| | | $n = 299$ | $\sum (x \cdot f) = 9,635.5$ |

$$\bar{x} = \frac{\sum (x \cdot f)}{n} = \frac{9,635.5}{299} \approx 32.2 \text{ years old}$$

**53.** Class width $= \dfrac{\text{Range}}{\text{Number of classes}} = \dfrac{297 - 127}{5} = 34 \Rightarrow 35$

| Class | Midpoint | Frequency, $f$ |
|-------|----------|----------------|
| 127-161 | 144 | 9 |
| 162-196 | 179 | 8 |
| 197-231 | 214 | 3 |
| 232-266 | 249 | 3 |
| 267-301 | 284 | 1 |
| | | $\sum f = 24$ |

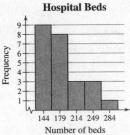

**Hospital Beds**

Shape: Positively skewed

**55.** Class width $= \dfrac{\text{Range}}{\text{Number of classes}} = \dfrac{76-62}{5} = 2.8 \Rightarrow 3$

| Class | Midpoint | Frequency, $f$ |
|-------|----------|----------------|
| 62-64 | 63 | 3 |
| 65-67 | 66 | 7 |
| 68-70 | 69 | 9 |
| 71-73 | 72 | 8 |
| 74-76 | 75 | 3 |
| | | $\sum f = 30$ |

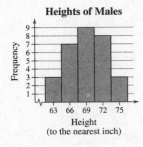

**Heights of Males**

Shape: Symmetric

**57.** (a) $\bar{x} = \dfrac{\sum x}{n} = \dfrac{36.03}{6} = 6.005$

    5.59  5.99  6.00  6.02  6.03  6.40

$$\text{median} = \dfrac{6.00+6.02}{2} = 6.01$$

(b) $\bar{x} = \dfrac{\sum x}{n} = \dfrac{35.67}{6} = 5.945$

    5.59  5.99  6.00  6.02  6.03  6.04

$$\text{median} = \dfrac{6.00+6.02}{2} = 6.01$$

(c) The mean was affected more.

**59.** Clusters around 16-21 and around 36

**61.** *Sample answer:* Option 2; The two clusters represent different types of vehicles which can be more meaningfully analyzed separately. For instance, suppose the mean gas mileage for cars is very far from the mean gas mileage for trucks, vans, and SUVs. Then, the mean gas mileage for all of the vehicles would be somewhere in the middle and would not accurately represent the gas mileages of either group of vehicles.

**63.** Car A

$$\bar{x} = \frac{\sum x}{n} = \frac{152}{5} = 30.4$$

28  28  **30**  32  34
$\nwarrow$ median = 30

mode = 28 (occurs 2 times)

Car B

$$\bar{x} = \frac{\sum x}{n} = \frac{151}{5} = 30.2$$

29  29  **31**  31  31
$\nwarrow$ median = 31

mode = 31 (occurs 3 times)

Car C

$$\bar{x} = \frac{\sum x}{n} = \frac{151}{5} = 30.2$$

28  29  **30**  32  32
$\nwarrow$ median = 30

mode = 32 (occurs 2 times)

(a)  Mean should be used because Car A has the highest mean of the three.
(b)  Median should be used because Car B has the highest median of the three.
(c)  Mode should be used because Car C has the highest mode of the three.

**65.** (a)  $\bar{x} = \frac{\sum x}{n} = \frac{1477}{30} \approx 49.2$

11 13 22 28 36 36 36 37 37 37 38 41 43 44 $\underbrace{46\ 47}$ 51 51 51 53 61 62 63 64

72 72 74 76 85 90                     median $= \frac{46 + 47}{2} = 46.5$

(b) Key: $3|6 = 36$

```
1|1  3
2|2  8           median
3|6  6  6  7 //  7  8
4|1  3  4  6 / 7
5|1  1  1  3           mean
6|1  2  3  4
7|2  2  4  6
8|5
9|0
```

(c) The distribution is positively skewed.

## 2.4 MEASURES OF VARIATION

### 2.4 Try It Yourself Solutions

**1a.** Min = 23, or \$23,000 and Max = 58, or \$58,000

  **b.** Range = Max − Min = 58 − 23 = 35, or \$35,000

  **c.** The range of the starting salaries for Corporation B is 35, or \$35,000. This is much larger than the range for Corporation A.

**2ab.** $\mu = 41.5$, or \$41,500

| Salary, $x$ | $x - \mu$ | $(x - \mu)^2$ |
|---|---|---|
| 23 | −18.5 | 342.25 |
| 29 | −12.5 | 156.25 |
| 32 | −9.5 | 90.25 |
| 40 | −1.5 | 2.25 |
| 41 | −0.5 | 0.25 |
| 41 | −0.5 | 0.25 |
| 49 | 7.5 | 56.25 |
| 50 | 8.5 | 72.25 |
| 52 | 10.5 | 110.25 |
| 58 | 16.5 | 272.25 |
| $\sum x = 415$ | $\sum (x - \mu) = 0$ | $\sum (x - \mu)^2 = 1102.5$ |

  **c.** $\sigma^2 = \dfrac{\sum (x - \mu)^2}{N} = \dfrac{1102.5}{10} \approx 110.3$

  **d.** $\sigma = \sqrt{\sigma^2} = \sqrt{\dfrac{1102.5}{10}} = 10.5$, or \$10,500

  **e.** The population variance is about 110.3 and the population standard deviation is 10.5, or \$10,500.

**3ab.** $\bar{x} = \dfrac{\sum x}{n} = \dfrac{316}{8} = 39.5$

| Time, $x$ | $x - \bar{x}$ | $(x - \bar{x})^2$ |
|:---:|:---:|:---:|
| 43 | 3.5 | 12.25 |
| 57 | 17.5 | 306.25 |
| 18 | -21.5 | 462.25 |
| 45 | 5.5 | 30.25 |
| 47 | 7.5 | 56.25 |
| 33 | -6.5 | 42.25 |
| 49 | 9.5 | 90.25 |
| 24 | -15.5 | 240.25 |
| $\sum x = 316$ | $\sum(x - \mu) = 0$ | $\sum(x - \mu)^2 = 1240$ |

$$SS_x = \sum(x - \bar{x})^2 = 1240$$

**b.** $s^2 = \dfrac{\sum(x - \bar{x})^2}{n-1} = \dfrac{1240}{7} \approx 177.1$

**c.** $s = \sqrt{s^2} = \sqrt{\dfrac{1240}{7}} \approx 13.3$

**4a.** Enter the data in a computer or a calculator.

**b.** $\bar{x} \approx 22.1, \quad s \approx 5.3$

**5a.** *Sample answer:* 7, 7, 7, 7, 7, 13, 13, 13, 13, 13

**b.**

| Salary, $x$ | $x - \mu$ | $(x - \mu)^2$ |
|:---:|:---:|:---:|
| 7 | −3 | 9 |
| 7 | −3 | 9 |
| 7 | −3 | 9 |
| 7 | −3 | 9 |
| 7 | −3 | 9 |
| 13 | 3 | 9 |
| 13 | 3 | 9 |
| 13 | 3 | 9 |
| 13 | 3 | 9 |
| 13 | 3 | 9 |
| $\sum x = 100$ | $\sum(x - \mu) = 0$ | $\sum(x - \mu)^2 = 90$ |

$$\mu = \dfrac{\sum x}{N} = \dfrac{100}{10} = 10$$

$$\sigma = \sqrt{\dfrac{\sum(x - \mu)^2}{N}} = \sqrt{\dfrac{90}{10}} = \sqrt{9} = 3$$

**6a.** $67.1 - 64.2 = 2.9 = 1$ standard deviation

**b.** 34%

**c.** Approximately 34% of women ages 20-29 are between 64.2 and 67.1 inches tall.

**7a.** $35.3 - 2(21.1) = -6.9$

Because $-6.9$ does not make sense for an age, use 0.

**b.** $35.3 + 2(21.1) = 77.5$

**c.** $1 - \dfrac{1}{k^2} = 1 - \dfrac{1}{(2)^2} = 1 - \dfrac{1}{4} = 0.75$

At least 75% of the data lie within 2 standard deviations of the mean. At least 75% of the population of Alaska is between 0 and 77.5 years old.

**8a.**

| $x$ | $f$ | $xf$ |
|---|---|---|
| 0 | 10 | 0 |
| 1 | 19 | 19 |
| 2 | 7 | 14 |
| 3 | 7 | 21 |
| 4 | 5 | 20 |
| 5 | 1 | 5 |
| 6 | 1 | 6 |
| | $n = 50$ | $\sum xf = 85$ |

**b.** $\overline{x} = \dfrac{\sum xf}{n} = \dfrac{85}{50} = 1.7$

**c.**

| $x - \overline{x}$ | $(x - \overline{x})^2$ | $(x - \overline{x})^2 f$ |
|---|---|---|
| $-1.7$ | 2.89 | 28.90 |
| $-0.7$ | 0.49 | 9.31 |
| 0.3 | 0.09 | 0.63 |
| 1.3 | 1.69 | 11.83 |
| 2.3 | 5.29 | 26.45 |
| 3.3 | 10.89 | 10.89 |
| 4.3 | 18.49 | 18.49 |
| | | $\sum (x - \overline{x})^2 f = 106.5$ |

**d.** $s = \sqrt{\dfrac{\sum (x - \overline{x})^2 f}{n-1}} = \sqrt{\dfrac{106.5}{49}} \approx 1.5$

**9a.**

| Class | $x$ | $f$ | $xf$ |
|---|---|---|---|
| 1-99 | 49.5 | 380 | 18,810 |
| 100-199 | 149.5 | 230 | 34,385 |
| 200-299 | 249.5 | 210 | 52,395 |
| 300-399 | 349.5 | 50 | 17,475 |
| 400-499 | 449.5 | 60 | 26,970 |
| 500+ | 650 | 70 | 45,500 |
| | | $n = 1000$ | $\sum xf = 195,535$ |

**b.** $\overline{x} = \dfrac{\sum xf}{n} = \dfrac{195,535}{1000} \approx 195.5$

c.

| $x - \bar{x}$ | $\left(x - \bar{x}\right)^2$ | $\left(x - \bar{x}\right)^2 f$ |
|---|---|---|
| −146.0 | 21,316 | 8,100,080 |
| −46.0 | 2116 | 486,680 |
| 54.0 | 2916 | 612,360 |
| 154.0 | 23,716 | 1,185,800 |
| 254.0 | 64,516 | 3,870,960 |
| 454.5 | 206,570.25 | 14,459,917.5 |
| | | $\sum\left(x - \bar{x}\right)^2 f = 28,715,797.5$ |

d.  $s = \sqrt{\dfrac{\sum\left(x - \bar{x}\right)^2 f}{n-1}} = \sqrt{\dfrac{28,715,797.5}{999}} \approx 169.5$

**10a.**   Los Angeles:  $\bar{x} \approx 31.0$ ,  $s \approx 12.6$

   Dallas/Fort Worth:  $\bar{x} \approx 22.1$ ,  $s \approx 5.3$

   **b.**   Los Angeles:  $CV = \dfrac{s}{\bar{x}} = \dfrac{12.6}{31.0} \cdot 100\% \approx 40.6\%$

   Dallas/Fort Worth:  $CV = \dfrac{s}{\bar{x}} = \dfrac{5.3}{22.1} \cdot 100\% \approx 24.0\%$

   **c.**   The office rental rates are more variable in Los Angeles than in Dallas/Fort Worth.

## 2.4 EXERCISE SOLUTIONS

1.   The range is the difference between the maximum and minimum values of a data set. The advantage of the range is that it is easy to calculate. The disadvantage is that it uses only two entries from the data set.

3.   The units of variance are squared. Its units are meaningless (example: dollars$^2$).  The units of standard deviation are the same as the data.

5.   When calculating the population standard deviation, you divide the sum of the squared deviations by $N$, then take the square root of that value. When calculating the sample standard deviation, you divide the sum of the squared deviations by $n-1$, then take the square root of that value.

7.   Similarity: Both estimate proportions of the data contained within $k$ standard deviations of the mean.
   Difference: The Empirical Rule assumes the distribution is approximately symmetric and bell-shaped. Chebychev's Theorem makes no such assumption.

9.   Range = Max − Min = 34 − 24 = 10

11.   (a)  Range = Max − Min = 38.5 − 20.7 = 17.8
   (b)  Range = Max − Min = 60.5 − 20.7 = 39.8

**13.** Range = Max − Min = 13 − 2 = 11

$$\mu = \frac{\sum x}{N} = \frac{121}{16} \approx 7.6$$

| $x$ | $x - \mu$ | $(x - \mu)^2$ |
|:---:|:---:|:---:|
| 13 | 5.4 | 29.16 |
| 10 | 2.4 | 5.76 |
| 12 | 4.4 | 19.36 |
| 11 | 3.4 | 11.56 |
| 7 | -0.6 | 0.36 |
| 8 | 0.4 | 0.16 |
| 6 | -1.6 | 2.56 |
| 6 | -1.6 | 2.56 |
| 10 | 2.4 | 5.76 |
| 7 | -0.6 | 0.36 |
| 12 | 4.4 | 19.36 |
| 4 | -3.6 | 12.96 |
| 6 | -1.6 | 2.56 |
| 5 | -2.6 | 6.76 |
| 2 | -5.6 | 31.36 |
| 2 | -5.6 | 31.36 |
| $\sum x = 121$ | $\sum (x - \mu) \approx 0$ | $\sum (x - \mu)^2 = 181.96$ |

$$\sigma^2 = \frac{\sum (x - \mu)^2}{N} = \frac{181.96}{16} \approx 11.4$$

$$\sigma = \sqrt{\frac{\sum (x - \mu)^2}{N}} = \sqrt{\frac{181.96}{16}} \approx 3.4$$

**15.** Range = Max − Min = 24 − 14 = 10

$$\bar{x} = \frac{\sum x}{n} = \frac{340}{20} = 17$$

| $x$ | $x - \bar{x}$ | $\left(x - \bar{x}\right)^2$ |
|:---:|:---:|:---:|
| 16 | -1 | 1 |
| 18 | 1 | 1 |
| 19 | 2 | 4 |
| 17 | 0 | 0 |
| 14 | -3 | 9 |
| 15 | -2 | 4 |
| 17 | 0 | 0 |
| 17 | 0 | 0 |
| 17 | 0 | 0 |
| 16 | -1 | 1 |
| 19 | 2 | 4 |
| 22 | 5 | 25 |
| 24 | 7 | 49 |
| 14 | -3 | 9 |
| 16 | -1 | 1 |
| 14 | -3 | 9 |
| 17 | 0 | 0 |
| 16 | -1 | 1 |
| 14 | -3 | 9 |
| 18 | 1 | 1 |
| $\sum x = 340$ | $\sum \left(x - \bar{x}\right) = 0$ | $\sum \left(x - \bar{x}\right)^2 = 128$ |

$$s^2 = \frac{\sum \left(x - \bar{x}\right)^2}{n-1} = \frac{128}{20-1} \approx 6.7$$

$$s = \sqrt{\frac{\sum \left(x - \bar{x}\right)^2}{n-1}} = \sqrt{\frac{128}{19}} \approx 2.6$$

**17.** The data set in (a) has a standard deviation of 24 and the data set in (b) has a standard deviation of 16 because the data in (a) have more variability.

**19.** Company B. An offer of $33,000 is two standard deviations from the mean of Company A's starting salaries, which makes it unlikely. The same offer is within one standard deviation of the mean of Company B's starting salaries, which makes the offer likely.

**21.** (a) Greatest sample standard deviation: (ii)
Data set (ii) has more entries that are farther away from the mean.
Least sample standard deviation: (iii)
Data set (iii) has more entries that are close to the mean.
(b) The three data sets have the same mean but have different standard deviations.

**23.** (a) Greatest sample standard deviation: (ii)

Data set (ii) has more entries that are farther away from the mean.

Least sample standard deviation: (iii)

Data set (iii) has more entries that are close to the mean.

(b) The three data sets have the same mean, median, and mode, but have different standard deviations.

**25.** *Sample answer:* 3,3,3,7,7,7

**27.** *Sample answer:* 9,9,9,9,9,9,9

**29.** $(63, 71) \rightarrow (67 - 1(4),\ 67 + 1(4)) \rightarrow (\bar{x} - s,\ \bar{x} + s)$

68% of the vehicles have speeds between 63 and 71 mph.

**31.** (a) $n = 75$; $68\%(75) = (0.68)(75) \approx 51$ vehicles have speeds between 63 and 71 mph.

(b) $n = 25$; $68\%(25) = (0.68)(25) \approx 17$ vehicles have speeds between 63 and 71 mph.

**33.** 78, 76, and 82 are unusual; 82 is very unusual because it is more than 3 standard deviations from the mean.

**35.** $(\bar{x} - 2s,\ \bar{x} + 2s) \rightarrow (0,\ 4)$ are 2 standard deviations from the mean.

$1 - \dfrac{1}{k^2} = 1 - \dfrac{1}{(2)^2} = 1 - \dfrac{1}{4} = 0.75 \Rightarrow$ At least 75% of the eruption times lie between 0 and 4.

If $n = 40$, at least $(0.75)(40) = 30$ households have between 0 and 4 pets.

**37.** $(\bar{x} - 2s,\ \bar{x} + 2s) \rightarrow (80,\ 96)$ are 2 standard deviations from the mean.

At least 75% of the test scores are from 80 to 96.

**39.**

| $x$ | $f$ | $xf$ | $x - \bar{x}$ | $(x - \bar{x})^2$ | $(x - \bar{x})^2 f$ |
|---|---|---|---|---|---|
| 0 | 3 | 0 | −1.74 | 3.0276 | 9.0828 |
| 1 | 15 | 15 | −0.74 | 0.5476 | 8.2140 |
| 2 | 24 | 48 | 0.26 | 0.0676 | 1.6224 |
| 3 | 8 | 24 | 1.26 | 1.5876 | 12.7008 |
| | $n = 50$ | $\sum xf = 87$ | | | $\sum (x - \bar{x})^2 f = 31.62$ |

$\bar{x} = \dfrac{\sum xf}{n} = \dfrac{87}{50} \approx 1.7$

$s = \sqrt{\dfrac{\sum (x - \bar{x})^2 f}{n - 1}} = \sqrt{\dfrac{31.62}{49}} \approx 0.8$

**41.**

| Class | Midpoint, $x$ | $f$ | $xf$ |
|-------|---------------|-----|------|
| 0-4   | 2  | 5  | 10  |
| 5-9   | 7  | 12 | 84  |
| 10-14 | 12 | 24 | 288 |
| 15-19 | 17 | 17 | 289 |
| 20-24 | 22 | 16 | 352 |
| 25-29 | 27 | 11 | 297 |
| 30+   | 32 | 5  | 160 |
|       |    | $n = 90$ | $\sum xf = 1480$ |

$$\overline{x} = \frac{\sum xf}{n} = \frac{1480}{90} \approx 16.4$$

| $x - \overline{x}$ | $\left(x - \overline{x}\right)^2$ | $\left(x - \overline{x}\right)^2 f$ |
|--------------------|-----------------------------------|-------------------------------------|
| -14.44 | 208.5136 | 1042.5680 |
| -9.44  | 89.1136  | 1069.3632 |
| -4.44  | 19.7136  | 473.1264  |
| 0.56   | 0.3136   | 5.3312    |
| 5.56   | 30.9136  | 494.6176  |
| 10.56  | 111.5136 | 1226.6496 |
| 15.56  | 242,1136 | 1210.5680 |
|        |          | $\sum \left(x - \overline{x}\right)^2 f = 5522.2240$ |

$$s = \sqrt{\frac{\sum \left(x - \overline{x}\right)^2 f}{n-1}} = \sqrt{\frac{5522.2240}{89}} \approx 7.9$$

**43.** Dallas: $\overline{x} = \dfrac{\sum x}{n} = \dfrac{442.8}{10} = 44.28$

| $x$ | $x - \overline{x}$ | $\left(x - \overline{x}\right)^2$ |
|-----|--------------------|-----------------------------------|
| 38.7 | −5.58 | 31.1364 |
| 39.9 | −4.38 | 19.1844 |
| 40.5 | −3.78 | 14.2884 |
| 41.6 | −2.68 | 7.1824  |
| 44.3 | 0.02  | 0.0004  |
| 44.7 | 0.42  | 0.1764  |
| 45.8 | 1.52  | 2.3104  |
| 47.8 | 3.52  | 12.3904 |
| 49.5 | 5.22  | 27.2484 |
| 50.0 | 5.72  | 32.7184 |
|      |       | $\sum \left(x - \overline{x}\right)^2 = 146.636$ |

$$s^2 = \frac{\sum \left(x - \overline{x}\right)^2}{n-1} = \frac{146.636}{9} \approx 16.29; \quad s = \sqrt{\frac{\sum \left(x - \overline{x}\right)^2}{n-1}} = \sqrt{\frac{146.636}{9}} \approx 4.04$$

$$CV = \frac{s}{\bar{x}} = \frac{4.04}{44.28} \cdot 100\% \approx 9.1\%$$

New York City: $\bar{x} = \frac{\sum x}{n} = \frac{509.2}{10} = 50.92$

| $x$ | $x - \bar{x}$ | $\left(x - \bar{x}\right)^2$ |
|---|---|---|
| 41.5 | −9.42 | 88.7364 |
| 42.3 | −8.62 | 74.3044 |
| 45.6 | −5.32 | 28.3024 |
| 47.2 | −3.72 | 13.8384 |
| 50.6 | −0.32 | 0.1024 |
| 51.0 | 0.08 | 0.0064 |
| 55.1 | 4.18 | 17.4724 |
| 57.6 | 6.68 | 44.6224 |
| 59.0 | 8.08 | 65.2864 |
| 59.3 | 8.38 | 70.2244 |
| | | $\sum\left(x - \bar{x}\right)^2 = 402.896$ |

$$s^2 = \frac{\sum\left(x - \bar{x}\right)^2}{n-1} = \frac{402.896}{9} \approx 44.77 \; ; \; s = \sqrt{\frac{\sum\left(x - \bar{x}\right)^2}{n-1}} = \sqrt{\frac{402.896}{9}} \approx 6.69$$

$$CV = \frac{s}{\bar{x}} = \frac{6.69}{50.92} \cdot 100\% \approx 13.1\%$$

Salaries for entry level accountants are more variable in New York City than in Dallas.

**45.** Ages: $\mu = \frac{\sum x}{N} = \frac{326}{12} \approx 27.2$

| $x$ | $x - \mu$ | $\left(x - \mu\right)^2$ |
|---|---|---|
| 24 | -3.17 | 10.0489 |
| 29 | 1.83 | 3.3489 |
| 37 | 9.83 | 96.6289 |
| 24 | -3.17 | 10.0489 |
| 26 | -1.17 | 1.3689 |
| 25 | -2.17 | 4.7089 |
| 24 | -3.17 | 10.0489 |
| 32 | 4.83 | 23.3289 |
| 22 | -5.17 | 26.7289 |
| 29 | 1.83 | 3.3489 |
| 23 | -4.17 | 17.3889 |
| 31 | 3.83 | 14.6689 |
| | | $\sum\left(x - \mu\right)^2 = 221.6668$ |

$$\sigma^2 = \frac{\sum\left(x - \mu\right)^2}{N} = \frac{221.6668}{12} \approx 18.5 \; ; \; \sigma = \sqrt{\frac{\sum\left(x - \bar{x}\right)^2}{N}} = \sqrt{\frac{221.6668}{12}} \approx 4.3$$

$$CV = \frac{\sigma}{\mu} = \frac{4.3}{27.2} \cdot 100\% \approx 15.8\%$$

Heights: $\mu = \dfrac{\sum x}{N} = \dfrac{896}{12} \approx 74.7$

| $x$ | $x - \mu$ | $(x - \mu)^2$ |
|---|---|---|
| 72 | -2.67 | 7.1289 |
| 76 | 1.33 | 1.7689 |
| 73 | -1.67 | 2.7889 |
| 73 | -1.67 | 2.7889 |
| 77 | 2.33 | 5.4289 |
| 76 | 1.33 | 1.7689 |
| 72 | -2.67 | 7.1289 |
| 74 | -0.67 | 0.4489 |
| 75 | 0.33 | 0.1089 |
| 75 | 0.33 | 0.1089 |
| 74 | -0.67 | 0.4489 |
| 79 | 4.33 | 18.7489 |
| | | $\sum (x - \mu)^2 = 48.6668$ |

$\sigma^2 = \dfrac{\sum (x - \mu)^2}{N} = \dfrac{48.6668}{12} \approx 4.1; \quad \sigma = \sqrt{\dfrac{\sum (x - \overline{x})^2}{N}} = \sqrt{\dfrac{48.6668}{12}} \approx 2.0$

$CV = \dfrac{\sigma}{\mu} = \dfrac{2.0}{74.7} \cdot 100\% \approx 2.7\%$

Ages are more variable than heights for all pitchers on the St. Louis Cardinals.

**47.** Team A: $\overline{x} = \dfrac{\sum x}{n} = \dfrac{2.991}{10} = 0.2991$

| $x$ | $x - \overline{x}$ | $(x - \overline{x})^2$ |
|---|---|---|
| 0.235 | −0.0641 | 0.00410881 |
| 0.256 | −0.0431 | 0.00185761 |
| 0.272 | −0.0271 | 0.00073441 |
| 0.295 | −0.0041 | 0.00001681 |
| 0.297 | −0.0021 | 0.00000441 |
| 0.297 | −0.0021 | 0.00000441 |
| 0.310 | 0.0109 | 0.00011881 |
| 0.320 | 0.0209 | 0.00043681 |
| 0.325 | 0.0259 | 0.00067081 |
| 0.384 | 0.0849 | 0.00720801 |
| | | $\sum (x - \overline{x})^2 = 0.0151609$ |

$s^2 = \dfrac{\sum (x - \overline{x})^2}{n - 1} = \dfrac{0.0151609}{9} \approx 0.0017; \quad s = \sqrt{\dfrac{\sum (x - \overline{x})^2}{n - 1}} = \sqrt{\dfrac{0.0151609}{9}} = 0.0410$

$CV = \dfrac{s}{\overline{x}} = \dfrac{.0410}{.2991} \cdot 100\% \approx 13.7\%$

Team B: $\bar{x} = \dfrac{\sum x}{n} = \dfrac{2.610}{10} = 0.261$

| $x$ | $x - \bar{x}$ | $\left(x - \bar{x}\right)^2$ |
|---|---|---|
| .204 | -0.057 | 0.003249 |
| .223 | -0.038 | 0.001444 |
| .226 | -0.035 | 0.001225 |
| .238 | -0.023 | 0.000529 |
| .256 | -0.005 | 0.000025 |
| .260 | -0.001 | 0.000001 |
| .292 | 0.031 | 0.000961 |
| .299 | 0.038 | 0.001444 |
| .300 | 0.039 | 0.001521 |
| .312 | 0.051 | 0.002601 |
| | | $\sum\left(x - \bar{x}\right)^2 = 0.013$ |

$$s^2 = \dfrac{\sum\left(x - \bar{x}\right)^2}{n-1} = \dfrac{0.013}{9} \approx 0.0014 \; ; \quad s = \sqrt{\dfrac{\sum\left(x - \bar{x}\right)^2}{n-1}} = \sqrt{\dfrac{0.013}{9}} \approx 0.0380$$

$$CV = \dfrac{s}{\bar{x}} = \dfrac{.0380}{.2610} \cdot 100\% \approx 14.6\%$$

Batting averages are slightly more variable on Team B than on Team A.

**49.** Ages:

| $x$ | $x^2$ |
|---|---|
| 16 | 256 |
| 18 | 324 |
| 19 | 361 |
| 17 | 289 |
| 14 | 196 |
| 15 | 225 |
| 17 | 289 |
| 17 | 289 |
| 17 | 289 |
| 16 | 256 |
| 19 | 361 |
| 22 | 484 |
| 24 | 576 |
| 14 | 196 |
| 16 | 256 |
| 14 | 196 |
| 17 | 289 |
| 16 | 256 |
| 14 | 196 |
| 18 | 324 |
| $\sum x = 340$ | $\sum x^2 = 5908$ |

$$s = \sqrt{\frac{\sum x^2 - \frac{(\sum x)^2}{n}}{n-1}} = \sqrt{\frac{5908 - \frac{(340)^2}{20}}{20-1}} = \sqrt{\frac{128}{19}} \approx 2.6$$

(b) The answer is the same as from Exercise 15.

**51.** (a) $\bar{x} \approx 41.7$     $s \approx 6.0$

    (b) $\bar{x} \approx 42.7$     $s \approx 6.0$

    (c) $\bar{x} \approx 39.7$     $s \approx 6.0$

    (d) By adding a constant $k$ to, or subtracting it from, each entry, the new sample mean will be $\bar{x} + k$, or $\bar{x} - k$, with the sample standard being unaffected.

**53.** $1 - \dfrac{1}{k^2} = 0.99 \Rightarrow 1 - 0.99 = \dfrac{1}{k^2} \Rightarrow k^2 = \dfrac{1}{0.01} \Rightarrow k = \sqrt{\dfrac{1}{0.01}} = 10$

At least 99% of the data in any data set lie within 10 standard deviations of the mean.

## 2.5 MEASURES OF POSITION

## 2.5 Try It Yourself Solutions

**1a.** 26, 31, 35, 37, 43, 43, 43, 44, 45, 47, 48, 48, 49, 50, 51, 51, 51, 51, 52, 54, 54, 54, 54, 55, 55, 55, 56, 57, 57, 57, 58, 58, 58, 58, 59, 59, 59, 62, 62, 63, 64, 65, 65, 65, 66, 66, 67, 67, 72, 86

  **b.** $Q_2 = 55$

  **c.** $Q_1 = 49$, $Q_3 = 62$

  **d.** About one fourth of the 50 most powerful women are 49 years old or younger; about one half are 55 years old or younger; and about three fourths of the 50 most powerful women are 62 years old or younger.

**2a.** Enter data

  **b.** $Q_1 = 23.5$, $Q_2 = 30$, $Q_3 = 45$

  **c.** About one-quarter of these universities charge tuition of $23,500 or less; about one-half charge $30,000 or less; and about three-quarters charge $45,000 or less.

**3a.** $Q_1 = 49$, $Q_3 = 62$

  **b.** IQR $= Q_3 - Q_1 = 62 - 49 = 13$

  **c.** $Q_1 - 1.5(IQR) = 49 - 1.5(13) = 29.5$;   $Q_3 + 1.5(IQR) = 62 + 1.5(13) = 81.5$ The age 26 is less than $Q_1 - 1.5(IQR)$ and the age 86 is greater than $Q_3 + 1.5(IQR)$.

  **d.** The ages of the 50 most powerful women in the middle portion of the data set vary by at most 13 years. The ages 62 and 86 are outliers.

**4a.** Min $= 26$, $Q_1 = 49$, $Q_2 = 55$, $Q_3 = 62$, Max $= 86$

  **bc.** **Ages of the 50 Most Powerful Women**

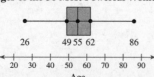

**d.** About 50% of the ages are between 49 and 62 years old. About 25% of the ages are less than 49 years old. About 25% of the ages are older than 62 years old.

**5a.** about 62

**b.** About 75% of the most powerful women are 62 years old or younger.

**6a.** 17,18,19,20,20,23,24,26,29,29,29,30,30,34,35,36,38,39,39,43,44,44,44,45,45

**b.** 7 data entries are less than 26

**c.** Percentile of $26 = \dfrac{\text{number of data entries less than 26}}{\text{total number of data entries}} \cdot 100 = \dfrac{7}{25} \cdot 100 = 28^{\text{th}}$ percentile

**d.** The tuition cost of $26,000 is greater than 28% of the other tuition costs.

**7a.** $\mu = 70,\ \sigma = 8$

$$x = 60:\ z = \frac{x - \mu}{\sigma} = \frac{60 - 70}{8} = -1.25$$

$$x = 71:\ z = \frac{x - \mu}{\sigma} = \frac{71 - 70}{8} = 0.125$$

$$x = 92:\ z = \frac{x - \mu}{\sigma} = \frac{92 - 70}{8} = 2.75$$

**b.** From the $z$-scores, the utility bill of $60 is 1.25 standard deviations below the mean, the bill of $71 is 0.125 standard deviation above the mean, and the bill of $92 is 2.75 standard deviations above the mean.

**8a.** 5 feet = 5(12) = 60 inches

**b.** Man: $z = \dfrac{x - \mu}{\sigma} = \dfrac{60 - 69.9}{3} = -3.3$; Woman: $z = \dfrac{x - \mu}{\sigma} = \dfrac{60 - 64.3}{2.6} \approx -1.7$

**c.** The $z$-score for the 5-foot-tall man is 3.3 standard deviations below the mean. This is an unusual height for a man. The $z$-score for the 5-foot-tall woman is 1.7 standard deviations below the mean. This is among the typical heights for a woman.

## 2.5 EXERCISE SOLUTIONS

**1.** The movie is shorter in length than 75% of the movies in the theater.

**3.** The student scored higher than 83% of the students who took the actuarial exam.

**5.** The interquartile range of a data set can be used to identify outliers because data values that are greater than $Q_3 + 1.5(\text{IQR})$ or less than $Q_1 - 1.5(\text{IQR})$ are considered outliers.

**7.** True

**9.** False. An outlier is any number above $Q_3 + 1.5(\text{IQR})$ or below $Q_1 - 1.5(\text{IQR})$.

**11. (a)** 51  54  56  **57**  59  60  60  **60**  60  62  63  **63**  63  65  80

$$\qquad\qquad\quad Q_1 \qquad\qquad Q_2 \qquad\qquad Q_3$$

**(b)** $\text{IQR} = Q_3 - Q_1 = 63 - 57 = 6$

(c) $Q_1 - 1.5(IQR) = 57 - 1.5(6) = 48$; $Q_3 + 1.5(IQR) = 63 + 1.5(6) = 72$. The date entry 80 is an outlier.

13. (a) 19  26  28  34  **36  36**  37  38  38  **40  41**  42  43  43  **45  48**  50  52  53  56

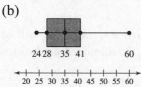

$Q_1 = 36$          $Q_2 = 40.5$          $Q_3 = 46.5$

(b) $IQR = Q_3 - Q_1 = 46.5 - 36 = 10.5$

(c) $Q_1 - 1.5(IQR) = 36 - 1.5(10.5) = 20.25$; $Q_3 + 1.5(IQR) = 46.5 + 1.5(10.5) = 62.25$. The data entry 19 is an outlier.

15. Min = 10, $Q_1 = 13$, $Q_2 = 15$, $Q_3 = 17$, Max = 20

17. (a) 24  26  27  **28**  30  32  35  **35**  36  39  39  **41**  50  51  60

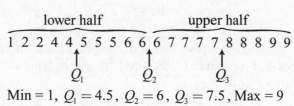

$Q_1$          $Q_2$          $Q_3$

Min = 24, $Q_1 = 28$, $Q_2 = 35$, $Q_3 = 41$, Max = 60

(b)

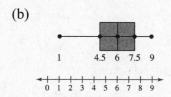

19. (a)

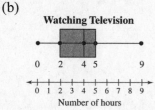

Min = 1, $Q_1 = 4.5$, $Q_2 = 6$, $Q_3 = 7.5$, Max = 9

(b)

21. None. The Data are not skewed or symmetric.

23. Skewed left. Most of the data lie to the right in the box-and-whisker plot.

25. (a) $Q_1 = 2$, $Q_2 = 4$, $Q_3 = 5$

(b)

**27.** (a) $Q_1 = 3$, $Q_2 = 3.85$, $Q_3 = 5.2$

   (b)

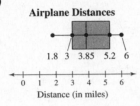

**Airplane Distances**

   1.8  3  3.85  5.2  6

   Distance (in miles)

**29.** (a) 5 (b) 50% (c) 25%

**31.** about 70 inches; About 60% of U.S. males ages 20-29 are shorter than 70 inches.

**33.** about 90th percentile; About 90% of U.S. males ages 20-29 are shorter than 73 inches.

**35.** Percentile of $40 = \dfrac{\text{number of data entries less than } 40}{\text{total number of data entries}} \cdot 100 = \dfrac{3}{30} \cdot 100 = 10^{\text{th}}$ percentile

**37.** $75^{\text{th}}$ percentile $= Q_3 = 56$; Ages over 56 are 57,57,61,61,65,66

**39.** $A \Rightarrow z = -1.43$
   $B \Rightarrow z = 0$
   $C \Rightarrow z = 2.14$
   The $z$-score 2.14 is unusual because it is so large.

**41.** (a) Bradley Wiggins: $x = 32 \Rightarrow z = \dfrac{x - \mu}{\sigma} = \dfrac{32 - 28.1}{3.4} \approx 1.15$

   (b) An age of 32 is about 1.15 standard deviations above the mean.
   (c) Not unusual.

**43.** (a) Cadel Evans: $x = 34 \Rightarrow z = \dfrac{x - \mu}{\sigma} = \dfrac{34 - 28.1}{3.4} \approx 1.74$

   (b) An age of 34 is about 1.74 standard deviations above the mean.
   (c) Not unusual.

**45.** (a) Firmin Lambot: $x = 36 \Rightarrow z = \dfrac{x - \mu}{\sigma} = \dfrac{36 - 28.1}{3.4} \approx 2.32$

   (b) An age of 36 is about 2.32 standard deviations above the mean.
   (c) Unusual.

**47.** (a) $x = 34{,}000 \Rightarrow z = \dfrac{x - \mu}{\sigma} = \dfrac{34{,}000 - 35{,}000}{2{,}250} \approx -0.44$

   $x = 37{,}000 \Rightarrow z = \dfrac{x - \mu}{\sigma} = \dfrac{37{,}000 - 35{,}000}{2{,}250} \approx 0.89$

   $x = 30{,}000 \Rightarrow z = \dfrac{x - \mu}{\sigma} = \dfrac{30{,}000 - 35{,}000}{2{,}250} \approx -2.22$

   The tire with a life span of 30,000 miles has an unusually short life span.

(b) $x = 30{,}500 \Rightarrow z = \dfrac{x - \mu}{\sigma} = \dfrac{30{,}500 - 35{,}000}{2{,}250} = -2 \Rightarrow$ about 2.5th percentile

$\quad\quad x = 37{,}250 \Rightarrow z = \dfrac{x - \mu}{\sigma} = \dfrac{37{,}250 - 35{,}000}{2{,}250} = 1 \Rightarrow$ about 84th percentile

$\quad\quad x = 35{,}000 \Rightarrow z = \dfrac{x - \mu}{\sigma} = \dfrac{35{,}000 - 35{,}000}{2{,}250} = 0 \Rightarrow$ about 50th percentile

49. Robert Duvall: $x = 53 \Rightarrow z = \dfrac{x - \mu}{\sigma} = \dfrac{53 - 44}{8.8} \approx 1.02$

    Jack Nicholson: $x = 46 \Rightarrow z = \dfrac{x - \mu}{\sigma} = \dfrac{46 - 50}{14.1} \approx -0.28$

    The age of Robert Duvall was about a standard deviation above the mean age of Best Actor winners, and the age of Jack Nicholson was less than 1 standard deviation below the mean age of Best Supporting Actor winners. Neither actor's age is unusual.

51. John Wayne: $x = 62 \Rightarrow z = \dfrac{x - \mu}{\sigma} = \dfrac{62 - 44}{8.8} \approx 2.05$

    Gig Young: $x = 56 \Rightarrow z = \dfrac{x - \mu}{\sigma} = \dfrac{56 - 50}{14.1} \approx 0.43$

    The age of John Wayne was more than 2 standard deviations above the mean age of Best Actor winners, which is unusual. The age of Gig Young was less than 1 standard deviation above the mean age of Best Supporting Actor winners, which is not unusual.

53.

    1  2  3  3  5  5  7  7  8  10

    $\quad\quad\uparrow\quad\quad\uparrow\quad\quad\uparrow$
    $\quad\quad Q_1\quad\ Q_2\quad\ Q_3$

    Midquartile $= \dfrac{Q_1 + Q_3}{2} = \dfrac{3 + 7}{2} = 5$

55. (a) The distribution of Concert 1 is symmetric. The distribution of Concert 2 is skewed right. Concert 1 has less variation.
    (b) Concert 2 is more likely to have outliers because it has more variation.
    (c) Concert 1, because 68% of the data should be between $\pm 16.3$ of the mean.
    (d) No, you do not know the number of songs played at either concert or the actual lengths of the songs.

57. (a)

    <u>lower half</u>      <u>upper half</u>

    2  7  8  9  9  10  10  11  11  12  12  13  15  16  24

    $\quad\quad\quad\quad\uparrow\quad\quad\quad\quad\uparrow\quad\quad\quad\quad\uparrow$
    $\quad\quad\quad\quad Q_1\quad\quad\quad\ Q_2\quad\quad\quad\ Q_3$

    $Q_1 = 9$, $Q_2 = 11$, $Q_3 = 13$

    $\text{IQR} = Q_3 - Q_1 = 13 - 9 = 4$

    $1.5 \times \text{IQR} = 6$

$$Q_1 - (1.5 \times \text{IQR}) = 9 - 6 = 3$$

$$Q_3 + (1.5 \times \text{IQR}) = 13 + 6 = 19$$

Any values less than 6 or greater than 19 are outliers. So, 2 and 24 are outliers.

(b)

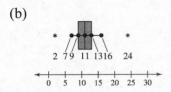

**59.** Answers will vary.

## CHAPTER 2 REVIEW EXERCISE SOLUTIONS

**1.** Class width = $\dfrac{\text{Max} - \text{Min}}{\text{Number of classes}} = \dfrac{30 - 8}{5} = 4.4 \Rightarrow 5$

| Class | Midpoint, $x$ | Boundaries | Frequency, $f$ | Relative frequency | Cumulative frequency |
|-------|---------------|------------|----------------|--------------------|----------------------|
| 8-12 | 10 | 7.5-12.5 | 2 | 0.10 | 2 |
| 13-17 | 15 | 12.5-17.5 | 10 | 0.50 | 12 |
| 18-22 | 20 | 17.5-22.5 | 5 | 0.25 | 17 |
| 23-27 | 25 | 22.5-27.5 | 1 | 0.05 | 18 |
| 28-32 | 30 | 27.5-32.5 | 2 | 0.10 | 20 |
| | | | $\sum f = 20$ | $\sum \dfrac{f}{n} = 1$ | |

**3.** Class width = $\dfrac{\text{Max} - \text{Min}}{\text{Number of classes}} = \dfrac{12.10 - 11.86}{7} \approx 0.03 \Rightarrow 0.04$

| Class | Midpoint | Frequency, $f$ | Relative frequency |
|-------|----------|----------------|--------------------|
| 11.86-11.89 | 11.875 | 3 | 0.125 |
| 11.90-11.93 | 11.915 | 5 | 0.208 |
| 11.94-11.97 | 11.955 | 8 | 0.333 |
| 11.98-12.01 | 11.995 | 7 | 0.292 |
| 12.02-12.05 | 12.035 | 0 | 0.000 |
| 12.06-12.09 | 12.075 | 0 | 0.000 |
| 12.10-12.13 | 12.115 | 1 | 0.042 |
| | | $\sum f = 24$ | $\sum \dfrac{f}{n} = 1$ |

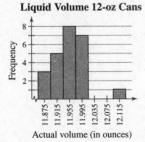

**Liquid Volume 12-oz Cans**

**5.** Class width $= \dfrac{\text{Max} - \text{Min}}{\text{Number of classes}} = \dfrac{166 - 79}{6} = 14.5 \Rightarrow 15$

| Class | Midpoint | Frequency, $f$ | Cumulative frequency |
|---|---|---|---|
| 79-93 | 86 | 9 | 9 |
| 94-108 | 101 | 12 | 21 |
| 109-123 | 116 | 5 | 26 |
| 124-138 | 131 | 3 | 29 |
| 139-153 | 146 | 2 | 31 |
| 154-168 | 161 | 1 | 32 |
| | | $\sum f = 32$ | |

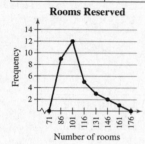

Rooms Reserved

**7.**

```
1 | 0  0                 Key: 1|0 = 10
2 | 0  0  2  5  5
3 | 0  3  4  5  5  8
4 | 1  2  4  4  7  8
5 | 2  3  3  7  9
6 | 1  1  5
7 | 1  5
8 | 9
```

*Sample answer:* Most cities have an air quality index from 20 to 59.

**9.**

| Location | Frequency | Relative frequency | Degrees |
|---|---|---|---|
| At home | 620 | 0.6139 | 221° |
| At friend's home | 110 | 0.1089 | 39° |
| At restaurant or bar | 50 | 0.0495 | 18° |
| Somewhere else | 100 | 0.0990 | 36° |
| Not sure | 130 | 0.1287 | 46° |
| | $\sum f = 1010$ | $\sum \dfrac{f}{n} = 1$ | |

**Location at Midnight on New Year's Day**
At home 61.39%
At friend's home 10.89%
Not sure 12.87%
At restaurant or bar 4.95%
Somewhere else 9.90%

*Sample answer:* Over half of the people surveyed will be at home on New Year's Day at midnight.

**11.**

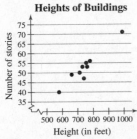

**Heights of Buildings**

*Sample answer:* The number of stories appears to increase with height.

**13.** $\bar{x} = \dfrac{\sum x}{n} = \dfrac{288}{10} = 28.8$

24.5  24.5  25.5  28.0  **28.5**  **29.5**  30.0  31.0  32.5  34.0

$\uparrow$

median $= 29$

Mode $= 24.5$ (occurs 2 times)

The mode does not represent the center of the data because 24.5 is the smallest number in the data set.

**15.**

| Source | Score, $x$ | Weight, $w$ | $x \cdot w$ |
|--------|-----------|-------------|-------------|
| Test 1 | 78 | 0.15 | 11.7 |
| Test 2 | 72 | 0.15 | 10.8 |
| Test 3 | 86 | 0.15 | 12.9 |
| Test 4 | 91 | 0.15 | 13.65 |
| Test 5 | 87 | 0.15 | 13.05 |
| Test 6 | 80 | 0.25 | 20 |
| | | $\sum w = 1$ | $\sum (x \cdot w) = 82.1$ |

$\bar{x} = \dfrac{\sum (x \cdot w)}{\sum w} = \dfrac{82.1}{1} = 82.1$

**17.**

| Midpoint, $x$ | Frequency, $f$ | $x \cdot f$ |
|---|---|---|
| 10 | 2 | 20 |
| 15 | 10 | 150 |
| 20 | 5 | 100 |
| 25 | 1 | 25 |
| 30 | 2 | 60 |
| | $n = 20$ | $\sum(x \cdot f) = 355$ |

$$\overline{x} = \frac{\sum(x \cdot f)}{n} = \frac{355}{20} \approx 17.8$$

**19.** Skewed right   **21.** Skewed left

**23.** Median, because the mean is to the left of the median in a skewed left distribution.

**25.** Range = Max − Min = 15 − 1 = 14

$$\mu = \frac{\sum x}{N} = \frac{96}{14} \approx 6.9$$

| $x$ | $x - \mu$ | $(x - \mu)^2$ |
|---|---|---|
| 4 | −2.9 | 8.41 |
| 2 | −4.9 | 24.01 |
| 9 | 2.1 | 4.41 |
| 12 | 5.1 | 26.01 |
| 15 | 8.1 | 65.61 |
| 3 | −3.9 | 15.21 |
| 6 | −0.9 | 0.81 |
| 8 | 1.1 | 1.21 |
| 1 | −5.9 | 34.81 |
| 4 | −2.9 | 8.41 |
| 14 | 7.1 | 50.41 |
| 12 | 5.1 | 26.01 |
| 3 | −3.9 | 15.21 |
| 3 | −3.9 | 15.21 |
| $\sum x = 96$ | $\sum(x - \mu) \approx 0$ | $\sum(x - \mu)^2 = 295.74$ |

$$\sigma^2 = \frac{\sum(x - \mu)^2}{N} = \frac{295.74}{14} \approx 21.1$$

$$\sigma = \sqrt{\frac{\sum(x - \mu)^2}{N}} = \sqrt{\frac{295.74}{14}} \approx 4.6$$

**27.** Range = Max − Min = \$6444 − \$4218 = \$2226

$$\bar{x} = \frac{\sum x}{n} = \frac{80{,}501}{15} \approx \$5366.73$$

| $x$ | $x - \bar{x}$ | $\left(x - \bar{x}\right)^2$ |
|---|---|---|
| 5306 | -60.73 | 3,688.1329 |
| 6444 | 1077.27 | 1,160,510.6529 |
| 5304 | -62.73 | 3,935.0529 |
| 4218 | -1148.73 | 1,319,580.6129 |
| 5159 | -207.73 | 43,151.7529 |
| 6342 | 975.27 | 951,151.5729 |
| 5713 | 346.27 | 119,902.9129 |
| 4859 | -507.73 | 257,789.7529 |
| 5365 | -1.73 | 2.9929 |
| 5078 | -288.73 | 83,365.0129 |
| 4334 | -1032.73 | 1,066,531.2529 |
| 5262 | -104.73 | 10,968.3729 |
| 5905 | 538.27 | 289,734.5929 |
| 6099 | 732.27 | 536,219.3529 |
| 5113 | -253.73 | 64,378.9129 |
| $\sum x = 80{,}501$ | $\sum\left(x - \bar{x}\right) \approx 0$ | $\sum\left(x - \bar{x}\right)^2 = 5{,}910{,}910.9335$ |

$$s^2 = \frac{\sum\left(x - \bar{x}\right)^2}{n-1} = \frac{5{,}910{,}910.9335}{14} \approx 422{,}207.92$$

$$s = \sqrt{\frac{\sum\left(x - \bar{x}\right)^2}{n-1}} = \sqrt{\frac{5{,}910{,}910.9335}{14}} \approx \$649.78$$

**29.** 99.7% of the distribution lies within 3 standard deviations of the mean.

$$\bar{x} - 3s = 70 - (3)(14.50) = 26.50$$

$$\bar{x} + 3s = 70 + (3)(14.50) = 113.50$$

99.7% of the distribution lies between \$26.50 and \$113.50.

**31.** $\left(\bar{x} - 2s,\ \bar{x} + 2s\right) \rightarrow \left(20,\ 52\right)$ are 2 standard deviations from the mean.

$$1 - \frac{1}{k^2} = 1 - \frac{1}{(2)^2} = 1 - \frac{1}{4} = 0.75$$

At least $(40)(0.75) = 30$ customers have a mean sale between \$20 and \$52.

**33.** $\bar{x} = \dfrac{\sum xf}{n} = \dfrac{99}{40} \approx 2.5$

| $x$ | $f$ | $xf$ | $x - \bar{x}$ | $\left(x - \bar{x}\right)^2$ | $\left(x - \bar{x}\right)^2 f$ |
|---|---|---|---|---|---|
| 0 | 1 | 0 | −2.5 | 6.25 | 6.25 |
| 1 | 8 | 8 | −1.5 | 2.25 | 18.00 |
| 2 | 13 | 26 | −0.5 | 0.25 | 3.25 |
| 3 | 10 | 30 | 0.5 | 0.25 | 2.50 |
| 4 | 5 | 20 | 1.5 | 2.25 | 11.25 |
| 5 | 3 | 15 | 2.5 | 6.25 | 18.75 |
| | $n = 40$ | $\sum xf = 99$ | | | $\sum \left(x - \bar{x}\right)^2 f = 60$ |

$$s = \sqrt{\dfrac{\sum \left(x - \bar{x}\right)^2 f}{n - 1}} = \sqrt{\dfrac{60}{39}} \approx 1.2$$

**35.** Freshmen: $\bar{x} = \dfrac{\sum x}{n} = \dfrac{23.1}{9} \approx 2.567$

| $x$ | $x - \bar{x}$ | $\left(x - \bar{x}\right)^2$ |
|---|---|---|
| 2.8 | 0.233 | 0.0543 |
| 1.8 | -0.767 | 0.5833 |
| 4.0 | 1.433 | 2.0535 |
| 3.8 | 1.233 | 1.5203 |
| 2.4 | -0.167 | 0.0279 |
| 2.0 | -0.567 | 0.3215 |
| 0.9 | -1.667 | 2.7789 |
| 3.6 | 1.033 | 1.0671 |
| 1.8 | -0.767 | 0.5883 |
| | | $\sum \left(x - \bar{x}\right)^2 = 9.0000$ |

$$s = \sqrt{\dfrac{\sum \left(x - \bar{x}\right)^2}{n - 1}} = \sqrt{\dfrac{9.0000}{8}} \approx 1.061$$

$$CV = \dfrac{s}{\bar{x}} = \dfrac{1.061}{2.567} \cdot 100\% \approx 41.3\%$$

Seniors: $\bar{x} = \dfrac{\sum x}{n} = \dfrac{26.6}{9} \approx 2.956$

| $x$ | $x - \bar{x}$ | $\left(x - \bar{x}\right)^2$ |
|---|---|---|
| 2.3 | -0.656 | 0.4303 |
| 3.3 | 0.344 | 0.1183 |
| 1.8 | -1.156 | 1.3363 |
| 4.0 | 1.044 | 1.0899 |
| 3.1 | 0.144 | 0.0207 |
| 2.7 | -0.256 | 0.0655 |
| 3.9 | 0.944 | 0.8911 |
| 2.6 | -0.356 | 0.1267 |
| 2.9 | -0.056 | 0.0031 |
| | | $\sum \left(x - \bar{x}\right)^2 = 4.0822$ |

$s = \sqrt{\dfrac{\sum \left(x - \bar{x}\right)^2}{n - 1}} = \sqrt{\dfrac{4.0822}{8}} \approx 0.714$

$CV = \dfrac{s}{\bar{x}} = \dfrac{0.714}{2.956} \cdot 100\% \approx 24.2\%$

Grade point averages are more variable for freshmen than seniors.

**37.**

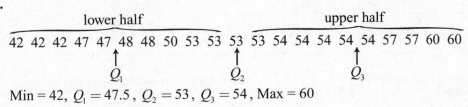

lower half          upper half

42 42 42 47 47 48 48 50 53 53   53   53 54 54 54 54 54 57 57 60 60

$Q_1$          $Q_2$          $Q_3$

Min = 42, $Q_1 = 47.5$, $Q_2 = 53$, $Q_3 = 54$, Max = 60

**39.**

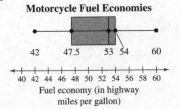

**Motorcycle Fuel Economies**

42   47.5   53 54   60

40 42 44 46 48 50 52 54 56 58 60
Fuel economy (in highway
miles per gallon)

**41.**

24.5 24.5 **25.5** 28.0 **28.5** **29.5** 30.0 **31.0** 32.5 34.0

$Q_1 = 25.5$      $Q_2 = 29$      $Q_3 = 31$

IQR $= Q_3 - Q_1 = 31 - 25.5 = 5.5$ inches

**43.** The 65th percentile means that 65% had a test grade of 75 or less. So, 35% scored higher than 75.

**45.** (a) $z = \dfrac{16{,}500 - 11{,}830}{2370} \approx 1.97$

(b) A towing capacity of 16,500 pounds is about 1.97 standard deviations above the mean.

(c) Not unusual

**47.** (a) $z = \dfrac{18{,}000 - 11{,}830}{2370} = 2.60$

(b) A towing capacity of 18,000 pounds is about 2.60 standard deviations above the mean.

(c) Unusual

## CHAPTER 2 QUIZ SOLUTIONS

**1.** (a) Class width $= \dfrac{\text{Max} - \text{Min}}{\text{Number of classes}} = \dfrac{157 - 101}{5} = 11.2 \Rightarrow 12$

| Class | Midpoint | Class boundaries | Frequency, $f$ | Relative frequency | Cumulative frequency |
|---|---|---|---|---|---|
| 101-112 | 106.5 | 100.5-112.5 | 3 | 0.12 | 3 |
| 113-124 | 118.5 | 112.5-124.5 | 11 | 0.44 | 14 |
| 125-136 | 130.5 | 124.5-136.5 | 7 | 0.28 | 21 |
| 137-148 | 142.5 | 136.5-148.5 | 2 | 0.08 | 23 |
| 149-160 | 154.5 | 148.5-160.5 | 2 | 0.08 | 25 |
|  |  |  | $\sum f = 25$ | $\sum \dfrac{f}{n} = 1$ |  |

(b) Frequency histogram and polygon

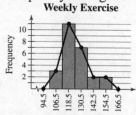

(c) Relative frequency histogram

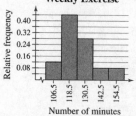

(d) Skewed right

(e)
```
10 | 1  8                    Key: 10|8 = 108
11 | 1  4  6  7  8  9  9
12 | 0  0  3  3  4  7  7  8
13 | 1  1  2  5  9  9
14 |
15 | 0  7
```

(f)

```
           lower half                                    upper half
101 108 111 114 116 117 118 119 119 120 120 123  123  124 127 127 128 131 131 132 135 139 139 150 157
                        ↑                          ↑                      ↑
                        Q₁                         Q₂                     Q₃
```

$\text{Min} = 101$, $Q_1 = 117.5$, $Q_2 = 123$, $Q_3 = 131.5$, $\text{Max} = 157$

**Weekly Exercise**

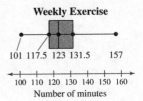

101  117.5  123  131.5    157

100 110 120 130 140 150 160
Number of minutes

(g)

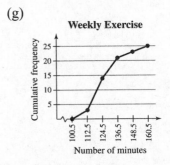

**Weekly Exercise**

Number of minutes

2. $\bar{x} = \dfrac{\sum xf}{n} = \dfrac{3130.5}{25} \approx 125.2$

| Midpoint, $x$ | Frequency, $f$ | $xf$ | $x-\bar{x}$ | $(x-\bar{x})^2$ | $(x-\bar{x})^2 f$ |
|---|---|---|---|---|---|
| 106.5 | 3 | 319.5 | −18.7 | 349.69 | 1049.07 |
| 118.5 | 11 | 1303.5 | −6.7 | 44.89 | 493.79 |
| 130.5 | 7 | 913.5 | 5.3 | 28.09 | 196.63 |
| 142.5 | 2 | 285.0 | 17.3 | 299.29 | 598.58 |
| 154.5 | 2 | 309.0 | 29.3 | 858.49 | 1716.98 |
| | $n=25$ | $\sum xf = 3130.5$ | | | $\sum(x-\bar{x})^2 f = 4055.05$ |

$s = \sqrt{\dfrac{\sum(x-\bar{x})^2 f}{n-1}} = \sqrt{\dfrac{4055.05}{24}} \approx 13.0$

3. (a)

| Category | Frequency | Relative frequency | Degrees |
|---|---|---|---|
| Clothing | 9.7 | 0.1187 | 43° |
| Footwear | 18.4 | 0.2252 | 81° |
| Equipment | 27.5 | 0.3366 | 121° |
| Rec. Transport | 26.1 | 0.3195 | 115° |
| | $n=81.7$ | $\sum\dfrac{f}{n}=1$ | |

**U.S. Sporting Goods**

Footwear 22.5%  Clothing 11.9%  Equipment 33.7%  Recreational transport 31.9%

(b)

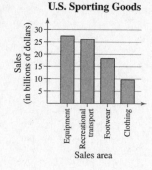

**4.** (a) $\bar{x} = \dfrac{\sum x}{n} = \dfrac{7413}{8} \approx 926.6$

619  621  842  **949**  **970**  1083  1135  1194

$$\text{median} = \dfrac{949 + 970}{2} = 959.5$$

mode = none

The mean best describes a typical salary because there are no outliers.

(b) Range = Max − Min = 1194 − 619 = 575

| $x$ | $x - \bar{x}$ | $\left(x - \bar{x}\right)^2$ |
|---|---|---|
| 949 | 22.4 | 501.76 |
| 621 | −305.6 | 93,391.36 |
| 1194 | 267.4 | 71,502.76 |
| 970 | 43.4 | 1883.56 |
| 1083 | 156.4 | 24,460.96 |
| 842 | −84.6 | 7157.16 |
| 619 | −307.6 | 94,617.76 |
| 1135 | 208.4 | 43,430.56 |
| | | $\sum\left(x - \bar{x}\right)^2 = 336{,}945.88$ |

$$s^2 = \dfrac{\sum\left(x - \bar{x}\right)^2}{n - 1} = \dfrac{336{,}945.88}{7} \approx 48{,}135.1$$

$$s = \sqrt{\dfrac{\sum\left(x - \bar{x}\right)^2}{n - 1}} = \sqrt{\dfrac{336{,}945.88}{7}} \approx 219.4$$

(c) $CV = \dfrac{s}{\bar{x}} = \dfrac{219.4}{926.6} \cdot 100\% \approx 23.7\%$

**5.** $\bar{x} - 2s = 155{,}000 - 2 \cdot 15{,}000 = \$125{,}000$

$\bar{x} + 2s = 155{,}000 + 2 \cdot 15{,}000 = \$185{,}000$

95% of the new home prices fall between \$125,000 and \$185,000.

**6.** (a) $x = 200{,}000:\ z = \dfrac{x - \bar{x}}{s} = \dfrac{200{,}000 - 155{,}000}{15{,}000} = 3.0 \Rightarrow$ unusual price

(b) $x = 55,000:$    $z = \dfrac{x - \bar{x}}{s} = \dfrac{55,000 - 155,000}{15,000} \approx -6.67 \Rightarrow$ very unusual price

(c) $x = 175,000:$    $z = \dfrac{x - \bar{x}}{s} = \dfrac{175,000 - 155,000}{15,000} \approx 1.33 \Rightarrow$ not unusual price

(d) $x = 122,000:$    $z = \dfrac{x - \bar{x}}{s} = \dfrac{122,000 - 155,000}{15,000} = -2.2 \Rightarrow$ unusual price

**7.** (a)
55 61 64 66 68 69 69 **72** 73 74 75 76 79 81 **81 83** 85 86 88 88 89 90 **93** 93 94 94 94 95 97 98

$Q_1 = 72$            $Q_2 = 82$            $Q_3 = 93$

Min $= 55$, $Q_1 = 72$, $Q_2 = 82$, $Q_3 = 93$, Max $= 98$

(b) IQR $= Q_3 - Q_1 = 93 - 72 = 21$

(c)

**Wins for Each Team**

55    72  82    93  98

50  55  60  65  70  75  80  85  90  95  100
Number of wins

## CUMULATIVE REVIEW FOR CHAPTERS 1 AND 2

**1.** Systematic sampling is used because every fortieth toothbrush from each assembly line is tested. It is possible for bias to enter into the sample if, for some reason, an assembly line makes a consistent error.

**3.**

**Reason for Baggage Delay**

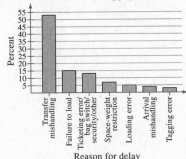

**5.** 10% is a statistic because it is describing a proportion within a sample of 1000 likely voters.

**7.** Population: Collection of opinions of all adults in the United States
Sample: Collection of opinions of the 1009 U.S. adults surveyed

**9.** Experiment. The study applies a treatment (stroke prevention devise) to the subjects.

**11.** Quantitative: The data are at the ratio level.

**13.** (a)

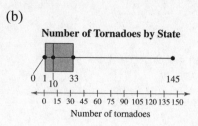

0 0 0 0 0 0 0 0 0 1 1   1 **1** 1 1 2 2 3 3 4 6 7 8 8 **10 10** 15 16 17 17
18 19 20 25 26 29 32 **33** 37 39 39 40 41 48 53 65 75 87 114 145

Min = 0, $Q_1 = 1$, $Q_2 = 10$, $Q_3 = 33$, Max = 145

(b)

**Number of Tornadoes by State**

0 1   33              145
  10

0  15  30  45  60  75  90  105 120 135 150
Number of tornadoes

(c) The distribution of the number of tornadoes is skewed right.

**15.** (a) $\bar{x} = \dfrac{49.4}{9} \approx 5.49$

3.4  3.9  4.2  4.6  ⑤.④  6.5  6.8  7.1  7.5

median = 5.4

mode = none
Both the mean and median accurately describe a typical American alligator tail length.
(Answers will vary.)

(b) Range – Max – Min – 7.5 – 3.4 = 4.1

| $x$ | $x - \bar{x}$ | $\left(x - \bar{x}\right)^2$ |
|-----|------|--------|
| 3.4 | −2.09 | 4.3681 |
| 3.9 | −1.59 | 2.5281 |
| 4.2 | −1.29 | 1.6641 |
| 4.6 | −0.89 | 0.7921 |
| 5.4 | −0.09 | 0.0081 |
| 6.5 | 1.01 | 1.0201 |
| 6.8 | 1.31 | 1.7161 |
| 7.1 | 1.61 | 2.5921 |
| 7.5 | 2.01 | 4.0401 |
|  |  | $\sum\left(x - \bar{x}\right)^2 = 18.7289$ |

$$s^2 = \frac{\sum\left(x - \bar{x}\right)^2}{n-1} = \frac{18.7289}{8} \approx 2.34$$

$$s = \sqrt{\frac{\sum\left(x - \bar{x}\right)^2}{n-1}} = \sqrt{\frac{18.7289}{8}} \approx 1.53$$

**17.** Class width $= \dfrac{\text{Max} - \text{Min}}{\text{Number of classes}} = \dfrac{65 - 0}{8} = 8.125 \Rightarrow 9$

| Class limits | Midpoint | Class boundaries | Frequency | Relative frequency | Cumulative frequency |
|---|---|---|---|---|---|
| 0-8 | 4 | −0.5-8.5 | 14 | 0.47 | 14 |
| 9-17 | 13 | 8.5-17.5 | 8 | 0.27 | 22 |
| 18-26 | 22 | 17.5-26.5 | 4 | 0.13 | 26 |
| 27-35 | 31 | 26.5-35.5 | 1 | 0.03 | 27 |
| 36-44 | 40 | 35.5-44.5 | 1 | 0.03 | 28 |
| 45-53 | 49 | 44.5-53.5 | 1 | 0.03 | 29 |
| 54-62 | 58 | 53.5-62.5 | 2 | 0.07 | 31 |
| 63-71 | 67 | 62.5-71.5 | 1 | 0.03 | 32 |
| | | | $\sum f = 30$ | $\sum \dfrac{f}{n} \approx 1$ | |

**19.**

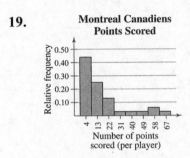

Class with greatest frequency: 0-8

Classes with least frequency: 27-35, 36-44, 45-53, and 63-71

# Probability

## 3.1 BASIC CONCEPTS OF PROBABILITY AND COUNTING

### 3.1 Try It Yourself Solutions

**1ab.** (1)                                              (2)

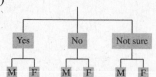     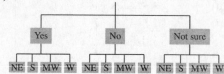

**c.** (1) 6 outcomes      (2) 12 outcomes

**d.** (1) Let $Y$ = Yes, $N$ = No, $NS$ = Not sure, $M$ = Male and $F$ = Female.
Sample space = {$YM, YF, NM, NF, NSM, NSF$}

(2) Let $Y$ = Yes, $N$ = No, $NS$ = Not sure, $NE$ = Northeast, $S$ = South, $MW$ = Midwest, and
$W$ = West
Sample space = {$YNE, YS, YMW, YW, NNE, NS, NMW, NW, NSNE, NSS, NSMW, NSW$}

**2a.** (1) Event $C$ has six outcomes: choosing the ages 18, 19, 20, 21, 22, and 23.
(2) Event $D$ has one outcome: choosing the age 20.

**b.** (1) The event is not a simple event because it consists of more than a single outcome.
(2) The event is a simple event because it consists of a single outcome.

**3a.** Manufacturer: 4        **b.** (4)(2)(5) = 40 ways        **c.** Tree Diagram for Car Selections
Size: 2
Color: 5

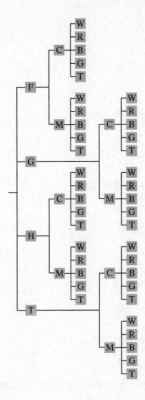

**4a.** (1) Each letter is an event (26 choices for each).
  (2) Each letter is an event (26, 25, 24, 23, 22, and 21 choices).
  (3) Each letter is an event (22, 26, 26, 26, 26, and 26 choices).

  **b.** (1) $26 \cdot 26 \cdot 26 \cdot 26 \cdot 26 \cdot 26 = 308,915,776$
  (2) $26 \cdot 25 \cdot 24 \cdot 23 \cdot 22 \cdot 21 = 165,765,600$
  (3) $22 \cdot 26 \cdot 26 \cdot 26 \cdot 26 \cdot 26 = 261,390,272$

**5a.** (1) 52  (2) 52  (3) 52

  **b.** (1) 1  (2) 13  (3) 52

  **c.** (1) $P(9 \text{ of clubs}) = \dfrac{1}{52} \approx 0.019$

  (2) $P(\text{heart}) = \dfrac{13}{52} = 0.25$

  (3) $P(\text{diamond, heart, club, or spade}) = \dfrac{52}{52} = 1$

**6a.** The event is "the next claim processed is fraudulent." The frequency is 4.

  **b.** Total Frequency $= 100$

  **c.** $P(\text{fraudulent claim}) = \dfrac{4}{100} = 0.04$

**7a.** Frequency $= 254$

**b.** Total of the Frequencies = 975

**c.** $P(\text{age 36 to 49}) = \dfrac{254}{975} \approx 0.261$

**8a.** The event is "salmon successfully passing through a dam on the Columbia River."

**b.** The probability is estimated from the results of an experiment.

**c.** Empirical probability

**9a.** $P(\text{age 18 to 22}) = \dfrac{156}{975} = 0.16$

**b.** $P(\text{age is not 18 to 22}) = 1 - \dfrac{156}{975} = \dfrac{819}{975} = 0.84$

**c.** $\dfrac{819}{975} = \dfrac{21}{25}$ or 0.84

**10a.** There are 5 outcomes in the event: {T1, T2, T3, T4, T5}.

**b.** $P(\text{tail and less than 6}) = \dfrac{5}{16} \approx 0.313$

**11a.** $10 \cdot 10 \cdot 10 \cdot 10 \cdot 10 \cdot 10 \cdot 10 = 10{,}000{,}000$

**b.** $\dfrac{1}{10{,}000{,}000}$

## 3.1 EXERCISE SOLUTIONS

1. An outcome is the result of a single trial in a probability experiment, whereas an event is a set of one or more outcomes.

3. It is impossible to have more than a 100% chance of rain.

5. The law of large numbers states that as an experiment is repeated over and over, the probabilities found in the experiment will approach the actual probabilities of the event. Examples will vary.

7. False. The event "tossing tails and rolling a 1 or a 3" is not a simple event because it consists of two possible outcomes and can be represented as $A = \{T1, T3\}$.

9. False. A probability of less than $\dfrac{1}{20} = 0.05$ indicates an unusual event.

11. b          13. c

**15.** {A, B, C, D, E, F, G, H, I, J, K, L, M, N, O, P, Q, R, S, T, U, V, W, X, Y, Z}; 26

**17.** {A♥, K♥, Q♥, J♥, 10♥, 9♥, 8♥, 7♥, 6♥, 5♥, 4♥, 3♥, 2♥,
    A♦, K♦, Q♦, J♦, 10♦, 9♦, 8♦, 7♦, 6♦, 5♦, 4♦, 3♦, 2♦,
    A♠, K♠, Q♠, J♠, 10♠, 9♠, 8♠, 7♠, 6♠, 5♠, 4♠, 3♠, 2♠,
    A♣, K♣, Q♣, J♣, 10♣, 9♣, 8♣, 7♣, 6♣, 5♣, 4♣, 3♣, 2♣}; 52

**19.**

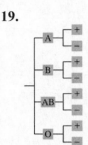

{(A, +), (B, +), (AB, +), (O, +), (A, –), (B, –), (AB, –), (O, –)}, where (A, +) represents positive Rh-factor with blood type A and (A, –) represents negative Rh-factor with blood type A; 8.

**21.** 1 outcome; simple event because it is an event that consists of a single outcome.

**23.** Ace = {Ace of hearts, Ace of spades, Ace of clubs, Ace of diamonds}; 4 outcomes
Not a simple event because it is an event that consists of more than a single outcome.

**25.** $(5)(10)(4) = 200$

**27.** $(9)(10)(10)(5) = 4500$

**29.** $P(A) = \dfrac{1}{12} \approx 0.083$

**31.** $P(C) = \dfrac{8}{12} = \dfrac{2}{3} \approx 0.667$

**33.** $P(E) = \dfrac{4}{12} \approx 0.333$

**35.** $P(\text{very prepared}) = \dfrac{259}{2163} \approx 0.120$

**37.** $P(18 \text{ to } 20) = \dfrac{4.2}{137.3} \approx 0.031$

**39.** $P(21 \text{ to } 24) = \dfrac{7.9}{137.3} \approx 0.058$

**41.** Empirical probability because company records were used to calculate the frequency of a washing machine breaking down.

**43.** Subjective probability because it is most likely based on an educated guess.

**45.** Classical probability because each outcome in the sample space is equally likely to occur.

**47.** $P(A) = 1 - P(A') = 1 - \dfrac{20}{193} \approx 1 - 0.104 \approx 0.896$

**49.** $P(C) = 1 - P(C') = 1 - \dfrac{38}{193} \approx 1 - 0.197 \approx 0.803$

**51-53.**

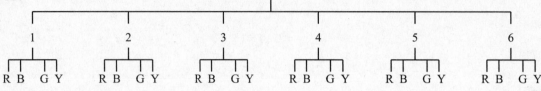

**51.** $P(A) = \dfrac{1}{24} \approx 0.042$ ; unusual      **53.** $P(C) = \dfrac{5}{24} \approx 0.208$ ; not unusual

**55.** (a) $10 \cdot 10 \cdot 10 = 1000$      (b) $\dfrac{1}{1000} = 0.001$      (c) $1 - \dfrac{1}{1000} = \dfrac{999}{1000} = 0.999$

**57.** {SSS, SSR, SRS, SRR, RSS, RSR, RRS, RRR}

**59.** {SSR, SRS, RSS}

**61.** $P(\text{voted in 2012 general election}) = \dfrac{3,896,846}{3,896,846 + 1,531,987} \approx 0.718$

**63.** $P(\text{doctorate}) = \dfrac{3}{91} \approx 0.033$

**65.** $P(\text{master's degree}) = \dfrac{25}{91} \approx 0.275$

**67.** Yes; The event in Exercise 37 can be considered unusual because its probability is less than or equal to 0.05.

**69.** (a) $P(\text{pink}) = \dfrac{2}{4} = 0.5$      (b) $P(\text{red}) = \dfrac{1}{4} = 0.25$      (c) $P(\text{white}) = \dfrac{1}{4} = 0.25$

**71.** $P(\text{service industry}) = \dfrac{115,675}{142,468} \approx 0.812$

**73.** $P(\text{not in service industry}) = 1 - P(\text{service industry}) \approx 1 - 0.812 = 0.188$

**75.** (a) $P(\text{at least 51}) = \dfrac{44}{120} \approx 0.367$

(b) $P(\text{between 20 and 30 inclusive}) = \dfrac{15}{120} = 0.125$

(c) $P(\text{more than } 72) = \dfrac{5}{120} \approx 0.042$ ; This event is unusual because its probability is 0.05 or less.

**77.** The probability of randomly choosing a tea drinker who does not have a college degree

**79.** (a)

| Sum | Outcomes | P(sum) | Probability |
|-----|----------|--------|-------------|
| 2 | (1, 1) | 1/36 | 0.028 |
| 3 | (1, 2), (2, 1) | 2/36 | 0.056 |
| 4 | (1, 3), (2, 2,), (3, 1) | 3/36 | 0.083 |
| 5 | (1, 4), (2, 3), (3, 2), (4, 1) | 4/36 | 0.111 |
| 6 | (1, 5), (2, 4), (3, 3), (4, 2), (5, 1) | 5/36 | 0.139 |
| 7 | (1, 6), (2, 5), (3, 4), (4, 3), (5, 2), (6, 1) | 6/36 | 0.167 |
| 8 | (2, 6), (3, 5), (4, 4), (5, 3), (6, 2) | 5/36 | 0.139 |
| 9 | (3, 6), (4, 5), (5, 4), (6, 3) | 4/36 | 0.111 |
| 10 | (4, 6), (5, 5), (6, 4) | 3/36 | 0.083 |
| 11 | (5, 6), (6, 5) | 2/36 | 0.056 |
| 12 | (6, 6) | 1/36 | 0.028 |

(b) Answers will vary.

(c) Answers will vary.

**81.** The first game; The probability of winning the second game is $\dfrac{1}{11} \approx 0.091$, which is less than $\dfrac{1}{10}$.

**83.** $13 : 39 = 1 : 3$

**85.** $p$ = number of successful outcomes

$q$ = number of unsuccessful outcomes

$$P(A) = \frac{\text{number of successful outcomes}}{\text{total number of outcomes}} = \frac{p}{p+q}$$

## 3.2 CONDITIONAL PROBABILITY AND THE MULTIPLICATION RULE

### 3.2 Try It Yourself Solutions

**1a.** (a) 30 and 102    (2) 11 and 50

**b.** $P(\text{does not have gene}) = \dfrac{30}{102} \approx 0.294$    (2) $P(\text{does not have gene} \mid \text{normal IQ}) = \dfrac{11}{50} = 0.22$

**2a.** (1) Yes    (2) No

**b.** (1) Dependent    (2) Independent

**3a.** (1) Independent    (2) Dependent

**b.** (1) Let $A$ = {first salmon swims successfully through the dam}

$B$ = {second salmon swims successfully through the dam}

$P(A \text{ and } B) = P(A) \cdot P(B) = (0.85) \cdot (0.85) \approx 0.723$

(2) Let $A$ = {selecting a heart}
  $B$ = {selecting a second heart}

$$P(A \text{ and } B) = P(A) \cdot P(B|A) = \left(\frac{13}{52}\right) \cdot \left(\frac{12}{51}\right) \approx 0.059$$

**4a.** (1) Find probability of the event      (2) Find probability of the event
  (3) Find probability of the compliment of the event

  **b.** (1) $P(3 \text{ surgeries are successful}) = (0.90) \cdot (0.90) \cdot (0.90) = 0.729$

    (2) $P(\text{none are successful}) = (0.10) \cdot (0.10) \cdot (0.10) = 0.001$

    (3) $P(\text{at least one rotator cuff surgery is successful}) = 1 - P(\text{none are successful})$
$$= 1 - 0.001 = 0.999$$

  **c.** (1) The event cannot be considered unusual because its probability is not less than or equal to 0.05.
  (2) The event can be considered unusual because its probability is less than or equal to 0.05.
  (3) The event cannot be considered unusual because its probability is not less than or equal to 0.05.

**5a.** (1),(2) $A$ = {is female}; $B$ = {works in health field}

  **b.** (1) $P(A \text{ and } B) = P(A) \cdot P(B|A) = (0.65) \cdot (0.25)$

    (2) $P(A \text{ and } B') = P(A) \cdot P(B'|A) = P(A) \cdot (1 - P(B|A)) = (0.65) \cdot (0.75)$

  **c.** (1) 0.163       (2) 0.488

  **d.** (1) and (2) The events are not unusual because their probabilitites are not less than or equal to 0.05.

## 3.2 EXERCISE SOLUTIONS

1. Two events are independent if the occurrence of one of the events does not affect the probability of the occurrence of the other event, whereas two events are dependent if the occurrence of one of the events does affect the probability of the occurrence of the other event.

3. The notation $P(B|A)$ means the probability of event $B$ occurring, given that event $A$ has occurred.

5. False. If two events are independent, $P(A|B) = P(A)$.

7. $A$ = {student is male}; $B$ = {student is nursing major};

    (a) $P(A|B) = \dfrac{94}{819} \approx 0.115$        (b) $P(B|A) = \dfrac{94}{1198} \approx 0.078$

9. Independent. The outcome of the first card drawn does not affect the outcome of the second card drawn.

**11.** Dependent. The outcome of a father having hazel eyes affects the outcome of a daughter having hazel eyes.

**13.** Dependent. The sum of the rolls depends on which numbers came up on the first and second rolls.

**15.** Events: moderate to severe sleep apnea, high blood pressure
Dependent. People with moderate to severe sleep apnea are more likely to have high blood pressure.

**16.** Events: stress, ulcers
Independent. Stress only irritates already existing ulcers.

**17.** Events: being around cell phones, developing cancer
Independent. Being around cell phones does not cause cancer.

**18.** Events: infection with dengue virus, mosquitoes being hungrier than usual
Dependent. Infection with dengue virus causes mosquitoes to be hungrier than usual.

**19.** Let $A = \{$card is a heart$\}$ and $B = \{$card is an ace$\}$.
$$P(A) = \frac{13}{52} = \frac{1}{4} \text{ and } P(B) = \frac{4}{52} = \frac{1}{13}. \text{ Thus, } P(A \text{ and } B) = \left(\frac{1}{4}\right)\cdot\left(\frac{1}{13}\right) \approx 0.019$$

**21.** Let $A = \{$woman carries mutation of BRCA gene$\}$ and $B = \{$woman develops breast cancer$\}$.
$$P(A) = \frac{1}{400} \text{ and } P(B \mid A) = \frac{6}{10} = \frac{3}{5}. \text{ Thus, } P(A \text{ and } B) = P(A)\cdot P(B\mid A) = \left(\frac{1}{400}\right)\cdot\left(\frac{3}{5}\right) \approx 0.002$$

**23.** Let $A = \{$first adult dines out at restaurant more than once a week$\}$ and $B = \{$second adult dines out at restaurant more than once a week$\}$.
$$P(A) = \frac{180}{1000} \text{ and } P(B \mid A) = \frac{179}{999}.$$

(a) $P(A \text{ and } B) = P(A)\cdot P(B\mid A) = \left(\frac{180}{1000}\right)\cdot\left(\frac{179}{999}\right) \approx 0.032$

(b) $P(A' \text{ and } B') = P(A')\cdot P(B'\mid A') = \left(\frac{820}{1000}\right)\cdot\left(\frac{819}{999}\right) \approx 0.672$

(c) $P(\text{at least 1 of 2 adults dines out more than once a week}) = 1 - P(A' \text{ and } B') \approx 1 - 0.672 \approx 0.328$

(d) The event in part (a) is unusual because its probability is less than or equal to 0.05.

**25.** Let $A = \{$first adult said Franklin Roosevelt best president$\}$ and $B = \{$second adult said Franklin Roosevelt best president$\}$.
$$P(A) = \frac{383}{2016} \text{ and } P(B \mid A) = \frac{382}{2015}.$$

(a) $P(A \text{ and } B) = P(A)\cdot P(B\mid A) = \left(\frac{383}{2016}\right)\cdot\left(\frac{382}{2015}\right) \approx 0.036$

(b) $P(A' \text{ and } B') = P(A')\cdot P(B'\mid A') = \left(\frac{1633}{2016}\right)\cdot\left(\frac{1632}{2015}\right) \approx 0.656$

(c) $P($at least 1 of 2 adults says Franklin Roosevelt best president$) = 1 - P(A'$ and $B') \approx 1 - 0.656 \approx 0.344$

(d) The event in part (a) is unusual because its probability is less than or equal to 0.05.

**27.** (a) $P($all five have B+$) = (0.09) \cdot (0.09) \cdot (0.09) \cdot (0.09) \cdot (0.09) \approx 0.000006$

(b) $P($none have B+$) = (0.91) \cdot (0.91) \cdot (0.91) \cdot (0.91) \cdot (0.91) \approx 0.624$

(c) $P($at least one has B+$) = 1 - P($none have B+$) \approx 1 - 0.624 = 0.376$

(d) The event in part (a) is unusual because its probability is less than or equal to 0.05.

**29.** Let $A = \{$ART procedure resulted in pregnancy$\}$ and $B = \{$ART pregnancy resulted in multiple birth$\}$.

$P(A) = 0.45$ and $P(B \mid A) = 0.24$

(a) $P(A$ and $B) = P(A) \cdot P(B \mid A) = (0.45) \cdot (0.24) = 0.108$

(b) $P(B' \mid A) = 1 - P(B \mid A) = 1 - 0.24 = 0.76$

(c) No, this is not unusual because the probability is not less than or equal to 0.05.

**31.** Let $A = \{$school library does not carry ebooks$\}$ and $B = \{$library does not plan to carry ebooks$\}$.

$P(A) = 0.56$ and $P(B \mid A) = 0.08$

$P(A$ and $B) = P(A) \cdot P(B \mid A) = (0.56) \cdot (0.08) = 0.045$

**33.** $P(A \mid B) = \dfrac{P(A) \cdot P(B \mid A)}{P(A) \cdot P(B \mid A) + P(A') \cdot P(B \mid A')}$

$$= \frac{\left(\dfrac{2}{3}\right) \cdot \left(\dfrac{1}{5}\right)}{\left(\dfrac{2}{3}\right) \cdot \left(\dfrac{1}{5}\right) + \left(\dfrac{1}{3}\right) \cdot \left(\dfrac{1}{2}\right)} = \frac{\dfrac{2}{15}}{\dfrac{3}{10}} = \frac{4}{9} \approx 0.444$$

**35.** $P(A \mid B) = \dfrac{P(A) \cdot P(B \mid A)}{P(A) \cdot P(B \mid A) + P(A') \cdot P(B \mid A')}$

$$= \frac{(0.25) \cdot (0.3)}{(0.25) \cdot (0.3) + (0.75) \cdot (0.5)} = \frac{0.075}{0.45} \approx 0.167$$

**37.** $P(A) = \dfrac{1}{200} = 0.005$, $P(B \mid A) = 0.8$, and $P(B \mid A') = 0.05$

(a) $P(A \mid B) = \dfrac{P(A) \cdot P(B \mid A)}{P(A) \cdot P(B \mid A) + P(A') \cdot P(B \mid A')}$

$$= \frac{(0.005) \cdot (0.8)}{(0.005) \cdot (0.8) + (0.995) \cdot (0.05)} = \frac{0.004}{0.05375} \approx 0.074$$

(b) $P(A'|B') = \dfrac{P(A') \cdot P(B'|A')}{P(A') \cdot P(B'|A') + P(A) \cdot P(B'|A)}$

$= \dfrac{(0.995) \cdot (0.95)}{(0.995) \cdot (0.95) + (0.005) \cdot (0.2)} = \dfrac{0.94525}{0.94625} \approx 0.999$

**39.** Let $A$ = {flight departs on time} and $B$ = {flight arrives on time}.

$P(A|B) = \dfrac{P(A \text{ and } B)}{P(B)} = \dfrac{0.83}{0.87} \approx 0.954$

## 3.3 THE ADDITION RULE

### 3.3 Try It Yourself Solutions

**1a.** (1) The events can occur at the same time.
    (2) The events cannot occur at the same time.

  **b.** (1) $A$ and $B$ are not mutually exclusive.
    (2) $A$ and $B$ are mutually exclusive.

**2a.** (1) Mutually exclusive         (2) Not mutually exclusive

  **b.** (1) Let $A$ = {6} and $B$ = {odd}.

    $P(A) = \dfrac{1}{6}$ and $P(B) = \dfrac{3}{6} = \dfrac{1}{2}$

    (2) Let $A$ = {face card} and $B$ = {heart}.

      $P(A) = \dfrac{12}{52}$, $P(B) = \dfrac{13}{52}$, and $P(A \text{ and } B) = \dfrac{3}{52}$

  **c.** (1) $P(A \text{ or } B) = P(A) + P(B) = \dfrac{1}{6} + \dfrac{1}{2} \approx 0.667$

    (2) $P(A \text{ or } B) = P(A) + P(B) - P(A \text{ and } B) = \dfrac{12}{52} + \dfrac{13}{52} - \dfrac{3}{52} \approx 0.423$

**3a.** Let $A$ = {sales between \$0 and \$24,999}
    Let $B$ = {sales between \$25,000 and \$49,999}.

  **b.** $A$ and $B$ cannot occur at the same time. So $A$ and $B$ are mutually exclusive.

  **c.** $P(A) = \dfrac{3}{36}$ and $P(B) = \dfrac{5}{36}$

  **d.** $P(A \text{ or } B) = P(A) + P(B) = \dfrac{3}{36} + \dfrac{5}{36} \approx 0.222$

**4a.** (1) Let $A$ = {type B} and $B$ = {type AB}.

(2) Let $A$ = {type O} and $B$ = {Rh-positive}.

**b.** (1) $A$ and $B$ cannot occur at the same time. So, $A$ and $B$ are mutually exclusive.

(2) $A$ and $B$ can occur at the same time. So, $A$ and $B$ are not mutually exclusive.

**c.** (1) $P(A) = \dfrac{45}{409}$ and $P(B) = \dfrac{16}{409}$

(2) $P(A) = \dfrac{184}{409}$, $P(B) = \dfrac{344}{409}$, and $P(A \text{ and } B) = \dfrac{156}{409}$

**d.** (1) $P(A \text{ or } B) = P(A) + P(B) = \dfrac{45}{409} + \dfrac{16}{409} \approx 0.149$

(2) $P(A \text{ or } B) = P(A) + P(B) - P(A \text{ and } B) = \dfrac{184}{409} + \dfrac{344}{409} - \dfrac{156}{409} \approx 0.910$

**5a.** Let $A$ = {linebacker} and $B$ = {quarterback}.

$$P(A \text{ or } B) = P(A) + P(B) = \frac{30}{253} + \frac{11}{253} \approx 0.162$$

**b.** $P$(not a linebacker or quarterback) = $1 - P(A \text{ or } B) \approx 1 - 0.162 = 0.838$

## 3.3 EXERCISE SOLUTIONS

**1.** $P(A \text{ and } B) = 0$ because $A$ and $B$ cannot occur at the same time.

**3.** True

**5.** False. The probability that event $A$ or event $B$ will occur is
$P(A \text{ or } B) = P(A) + P(B) - P(A \text{ and } B)$.

**7.** Not mutually exclusive. A presidential candidate can lose the popular vote and win the election..

**9.** Not mutually exclusive. A public school teacher can be female and be 25 years old.

**11.** Mutually exclusive. A person cannot be both a Republican and a Democrat.

**13.** Let $A$ = {biology major} and $B$ = {male}.

$P(A) = \dfrac{10}{32}$, $P(B) = \dfrac{14}{32}$, and $P(B \mid A) = \dfrac{4}{10}$.

Thus, $P(A \text{ and } B) = P(A) \cdot P(B \mid A) = \dfrac{10}{32} \cdot \dfrac{4}{10} = \dfrac{1}{8}$.

$P(A \text{ or } B) = P(A) + P(B) - P(A \text{ and } B) = \dfrac{10}{32} + \dfrac{14}{32} - \dfrac{1}{8} = \dfrac{20}{32} = 0.625$

**15.** Let $A = \{\text{puncture}\}$ and $B = \{\text{smashed corner}\}$.
$P(A) = 0.05$, $P(B) = 0.08$, and $P(A \text{ and } B) = 0.004$.
$P(A \text{ or } B) = P(A) + P(B) - P(A \text{ and } B) = 0.05 + 0.08 - 0.004 = 0.126$

**17.** (a) $P(\text{club or } 3) = P(\text{club}) + P(3) - P(\text{club and } 3) = \dfrac{13}{52} + \dfrac{4}{52} - \dfrac{1}{52} \approx 0.308$

   (b) $P(\text{red or king}) = P(\text{red}) + P(\text{king}) - P(\text{red and king}) = \dfrac{26}{52} + \dfrac{4}{52} - \dfrac{2}{52} \approx 0.538$

   (c) $P(9 \text{ or face card}) = P(9) + P(\text{face card}) - P(9 \text{ and face card}) = \dfrac{4}{52} + \dfrac{12}{52} - 0 \approx 0.308$

**19.** (a) $P(\text{under } 5) = 0.066$

   (b) $P(45+) = P(45-64) + P(65-74)6 + P(75+) = 0.248 + 0.096 + 0.066 = 0.41$

   (c) $P(\text{not } 65+) = 1 - P(65+) = 1 - P(65-74) - P(74+) = 1 - 0.096 - 0.066 = 0.838$

   (d) $P(\text{between 20 and 34}) = P(20-24) + P(25-34) = 0.064 + 0.134 = 0.198$

**21.** (a) $P(\text{not } A) = 1 - P(A) = 1 - \dfrac{52}{1026} = \dfrac{974}{1026} \approx 0.949$

   (b) $P(\text{better than } D) = P(A) + P(B) + P(C) = \dfrac{52}{1026} + \dfrac{241}{1026} + \dfrac{335}{1026} = \dfrac{628}{1026} \approx 0.612$

   (c) $P(D \text{ or } F) = P(D) + P(F) = \dfrac{272}{1026} + \dfrac{126}{1026} = \dfrac{398}{1026} \approx 0.388$

   (d) $P(A \text{ or } B) = P(A) + P(B) = \dfrac{52}{1026} + \dfrac{241}{1026} = \dfrac{293}{1026} \approx 0.286$

**23.** Let $A = \{\text{male}\}$; $B = \{\text{nursing major}\}$

   (a) $P(A \text{ or } B) = P(A) + P(B) - P(A \text{ and } B) = \dfrac{1198}{3605} + \dfrac{819}{3605} - \dfrac{94}{3605} \approx 0.533$

   (b) $P(A' \text{ or } B') = P(A') + P(B') - P(A' \text{ and } B') = \dfrac{2407}{3605} + \dfrac{2786}{3605} - \dfrac{1682}{3605} \approx 0.974$

   (c) $P(A \text{ or } B) \approx 0.533$

**25.** $A = \{\text{frequently}\}$; $B = \{\text{occasionally}\}$; $C = \{\text{not at all}\}$; $D = \{\text{male}\}$

   (a) $P(A \text{ or } B) = \dfrac{428}{2850} + \dfrac{886}{2850} \approx 0.461$

   (b) $P(D' \text{ or } C) = P(D') + P(C) - P(D' \text{ and } C) = \dfrac{1378}{2850} + \dfrac{1536}{2850} - \dfrac{741}{2850} \approx 0.762$

   (c) $P(D \text{ or } A) = P(D) + P(A) - P(D \text{ and } A) = \dfrac{1472}{2850} + \dfrac{428}{2850} - \dfrac{221}{2850} \approx 0.589$

   (d) $P(D' \text{ or } A') = P(D') + P(A') - P(D' \text{ and } A') = \dfrac{1378}{2850} + \dfrac{886 + 1536}{2850} - \dfrac{430 + 741}{2850} \approx 0.922$

**27.** No. If two events $A$ and $B$ are independent, then $P(A \text{ and } B) = P(A) \cdot P(B)$. If two events are mutually exclusive, $P(A \text{ and } B) = 0$. The only scenario when two events can be independent and mutually exclusive is if $P(A) = 0$ or $P(B) = 0$.

**29.** $P(A \text{ or } B \text{ or } C) = P(A) + P(B) + P(C) - P(A \text{ and } B) - P(A \text{ and } C)$
$$- P(B \text{ and } C) + P(A \text{ and } B \text{ and } C)$$
$$= 0.38 + 0.26 + 0.14 - 0.12 - 0.03 - 0.09 + 0.01 = 0.55$$

## 3.4 ADDITIONAL TOPICS IN PROBABILITY AND COUNTING

### 3.4 Try It Yourself Solutions

**1a.** $n = 8$ teams      **b.** $8! = 40,320$

**2a.** $n = 8, r = 3$

  **b.** $\dfrac{8!}{(8-3)!} = \dfrac{8!}{5!} = \dfrac{8 \cdot 7 \cdot 6 \cdot 5 \cdot 4 \cdot 3 \cdot 2 \cdot 1}{5 \cdot 4 \cdot 3 \cdot 2 \cdot 1} = 8 \cdot 7 \cdot 6 = 336$

  **c.** There are 336 possible ways that the subject can pick a first, second, and third activity.

**3a.** $n = 12, r = 4$

  **b.** $_{12}P_4 = \dfrac{12!}{(12-4)!} = \dfrac{12!}{8!} = 12 \cdot 11 \cdot 10 \cdot 9 = 11,880$

**4a.** $n = 20$

  **b.** Oak tree, maple tree, poplar tree

  **c.** $n_1 = 6$, $n_2 = 9$, $n_3 = 5$

  **d.** $\dfrac{n!}{n_1! \, n_2! \, n_3!} = \dfrac{20!}{6! \, 9! \, 5!} = 77,597,520$

**5a.** $n = 20, r = 3$

  **b.** $_{20}C_3 = \dfrac{20!}{(20-3)! \, 3!} = \dfrac{20!}{17! \, 3!} = \dfrac{20 \cdot 19 \cdot 18 \cdot 17!}{17! \, 3!} = 1140$

  **c.** There are 1140 different possible three-person committees that can be selected from 20 employees.

**6a.** $_{20}P_2 = \dfrac{20!}{(20-2)!} = \dfrac{20!}{18!} = \dfrac{20 \cdot 19 \cdot 18!}{18!} = 20 \cdot 19 = 380$

**b.** $P(\text{selecting the two members}) = \dfrac{1}{380} \approx 0.003$

**7a.** $_{15}C_5 = \dfrac{15!}{5!(15-5)!} = 3003$ **b.** $_{54}C_5 = \dfrac{54!}{5!(54-5)!} = 3{,}162{,}510$ **c.** $\dfrac{3003}{3{,}162{,}510} \approx 0.0009$

**8a.** $_5C_3 \cdot {_7C_0} = \dfrac{5!}{3!2!} \cdot \dfrac{7!}{0!7!} = 10 \cdot 1 = 10$ **b.** $_{12}C_3 = \dfrac{12!}{3!9!} = 220$ **c.** $\dfrac{10}{220} \approx 0.045$

## 3.4 EXERCISE SOLUTIONS

1. The number of ordered arrangements of $n$ objects taken $r$ at a time.
   *Sample answer*: An example of a permutation is the number of seating arrangements of you and three friends.

3. False. A permutation is an ordered arrangement of objects.

5. True

7. $_9P_5 = \dfrac{9!}{(9-5)!} = \dfrac{9!}{4!} = 9 \cdot 8 \cdot 7 \cdot 6 \cdot 5 = 15{,}120$

9. $_8C_3 = \dfrac{8!}{(8-3)!3!} = \dfrac{8!}{5!3!} = \dfrac{8 \cdot 7 \cdot 6 \cdot 5!}{5! \, 3!} = 56$

11. $\dfrac{_8C_4}{_{12}C_6} = \dfrac{\dfrac{8!}{(8-4)!4!}}{\dfrac{12!}{(12-6)!6!}} = \dfrac{\dfrac{8!}{4!4!}}{\dfrac{12!}{6!6!}} = \dfrac{70}{924} \approx 0.076$

13. $\dfrac{_6P_2}{_{11}P_3} = \dfrac{\dfrac{6!}{(6-2)!}}{\dfrac{11!}{(11-3)!}} = \dfrac{\dfrac{6!}{4!}}{\dfrac{11!}{8!}} = \dfrac{6 \cdot 5}{11 \cdot 10 \cdot 9} \approx 0.030$

15. Permutation. The order of the 8 cars in line matters.

17. Combination. The order does not matter because the position of one captain is the same as the other.

19. $7! = 5040$    21. $6! = 720$

23. $_{50}P_3 = \dfrac{50!}{(50-3)!} = \dfrac{50!}{47!} = 117{,}600$    25. $_{24}P_6 = \dfrac{24!}{(24-6)!} = \dfrac{24!}{18!} = 96{,}909{,}120$

**27.** $\dfrac{22!}{4!10!8!} = 320{,}089{,}770$

**29.** 3 $S$'s, 3 $T$'s, 1 $A$, 2 $I$'s, 1 $C$

$$\dfrac{10!}{3!3!1!2!1!} = 50{,}400$$

**31.** $_{20}C_4 = 4845$

**33.** $_{30}C_5 = \dfrac{30!}{(30-5)!5!} = 142{,}506$

**35.** $10 \cdot 8 \cdot {}_{13}C_2 = 10 \cdot 8 \cdot \dfrac{13!}{(13-2)!2!} = 6240$

**37.** $_5C_1 \cdot {}_{75}C_5 = \dfrac{5!}{1!4!} \cdot \dfrac{75!}{70!5!} = 86{,}296{,}950$

**39.** $\dfrac{1}{_6P_2} = \dfrac{1}{30} \approx 0.033$

**41.** $\dfrac{1}{_{12}C_3} = \dfrac{1}{220} \approx 0.005$

**43.** (a) $\dfrac{_{15}C_3}{_{56}C_3} = \dfrac{455}{27{,}720} \approx 0.016$

(b) $\dfrac{_{41}C_3}{_{56}C_3} = \dfrac{10{,}660}{27{,}720} \approx 0.385$

**45.** $(7\%)(1200) = (0.07)(1200) = 84$ of the 1200 rate financial shape as excellent.

$$P\left(\text{all four rate excellent}\right) = \dfrac{_{84}C_4}{_{1200}C_4} = \dfrac{1{,}929{,}501}{85{,}968{,}659{,}700} \approx 0.00002$$

**47.** $(38\%)(500) = (0.38)(500) = 190$ of the 500 rate financial shape as fair $\Rightarrow 500 - 190 = 310$ rate financial shape as not fair.

$$P\left(\text{none of 80 rate fair}\right) = \dfrac{_{310}C_{80}}{_{500}C_{80}} \approx 2.70 \times 10^{-19}$$

**49.** (a) $_{40}C_5 = 658{,}008$

(b) $P(\text{win}) = \dfrac{1}{658{,}008} \approx 0.0000015$

**51.** $\dfrac{\left(_{24}C_6\right)\left(_{17}C_2\right)}{_{41}C_8} = \dfrac{(134{,}596)(136)}{95{,}548{,}245} \approx 0.192$

**53.** $\dfrac{\left(_{24}C_4\right)\left(_{17}C_4\right)}{_{41}C_8} = \dfrac{(10{,}626)(2380)}{95{,}548{,}245} \approx 0.265$

**55.** $\dfrac{_8C_3}{_{10}C_3} + \dfrac{\left(_8C_2\right)\left(_2C_1\right)}{_{10}C_3} = \dfrac{56}{120} + \dfrac{(28)(2)}{120} = \dfrac{112}{120} \approx 0.933$

**57.** $_4C_2\left(\dfrac{\left(_2C_2\right)\left(_2C_2\right)\left(_2C_0\right)\left(_2C_0\right)}{_8C_4}\right) = 6 \cdot \dfrac{(1)(1)(1)(1)}{70} \approx 0.086$

**59.** $\dfrac{\left(_{13}C_2\right)\left(_{13}C_1\right)\left(_{13}C_1\right)\left(_{13}C_1\right)}{_{52}C_5} = \dfrac{(78)(13)(13)(13)}{2{,}598{,}960} \approx 0.066$

**61.** $\left(_{13}C_2\right)\left(\dfrac{\left(_4C_3\right)\left(_4C_2\right)}{_{52}C_5}\right) = 78\left(\dfrac{(4)(6)}{2{,}598{,}960}\right) \approx 0.001$

## CHAPTER 3 REVIEW EXERCISE SOLUTIONS

**1.** Sample space:
{HHHH, HHHT, HHTH, HHTT, HTHH, HTHT, HTTH, HTTT, THHH, THHT, THTH, THTT, TTHH, TTHT, TTTH, TTTT}

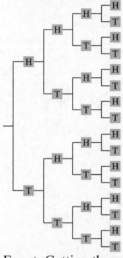

Event: Getting three heads
{HHHT, HHTH, HTHH, THHH}
There are 4 outcomes.

**3.** Sample space: {January, February, March, April, May, June, July, August, September, October, November, December}
Event: {January, June, July}
There are 3 outcomes.

**5.** $(7)(4)(3) = 84$

**7.** Empirical probability because prior counts were used to calculate the frequency of a part being defective.

**9.** Subjective probability because it is based on opinion.

**11.** Classical probability because all the outcomes in the event and the sample space can be counted.

**13.** $P(\text{at least } 10) = 0.096 + 0.100 + 0.224 = 0.42$

**15.** $\dfrac{1}{(8)(10)(10)(10)(10)(10)(10)} = 1.25 \times 10^{-7}$

**17.** $P(\text{failed}|\text{first}) = \dfrac{2058}{6485} \approx 0.317$

**19.** Independent. The outcomes of the first four coin tosses do not affect the outcome of the fifth coin toss.

**21.** Dependent. The outcome of getting high grades affects the outcome of being awarded an academic scholarship.

**23.** $P(\text{correct toothpaste and correct dental rinse}) = P(\text{correct toothpaste}) \cdot P(\text{correct dental rinse})$
$$= \dfrac{1}{8} \cdot \dfrac{1}{5} = \dfrac{1}{40} = 0.025$$

The event is unusual because its probability is less than or equal to 0.05.

**25.** Mutually exclusive. A jelly bean cannot be both completely red and completely yellow.

**27.** Mutually exclusive. A person cannot legally be registered to vote in more than one state.

**29.** $P(\text{home or work}) = P(\text{work}) + P(\text{home}) - P(\text{home and work}) = 0.74 + 0.88 - 0.72 = 0.9$

**31.** $P(4-8 \text{ or club}) = P(4-8) + P(\text{club}) - P(4-8 \text{ and club}) = \dfrac{20}{52} + \dfrac{13}{52} - \dfrac{5}{52} \approx 0.538$

**33.** $P(\text{odd or less than 4}) = P(\text{odd}) + P(\text{less than 4}) - P(\text{odd and less than 4}) = \dfrac{6}{12} + \dfrac{3}{12} - \dfrac{2}{12} \approx 0.583$

**35.** $P(500 \text{ or more}) = P(500-999) + P(1000 \text{ or more}) = 0.14 + 0.037 = 0.177$

**37.** $P(\text{Catholic or Muslim}) = \dfrac{76}{326} + \dfrac{2}{326} \approx 0.239$

**39.** $P(\text{not Protestant/Other Christian}) = 1 - P(\text{Protestant/Other Christian}) = 1 - \dfrac{169}{326} \approx 0.482$

**41.** $_{11}P_2 = \dfrac{11!}{(11-2)!} = \dfrac{11!}{9!} = 11 \cdot 10 = 110$

**43.** $_7C_4 = \dfrac{7!}{(7-4)!4!} = \dfrac{7!}{3!4!} = 35$

**45.** Order is important: $_{15}P_3 = \dfrac{15!}{(15-3)!} = \dfrac{15!}{12!} = 2730$

**47.** Order is not important: $_{17}C_4 = \dfrac{17!}{(17-4)!4!} = 2380$

**49.** $P(3 \text{ kings and } 2 \text{ queens}) = \dfrac{\left(_4C_3\right) \cdot \left(_4C_2\right)}{_{52}C_5} = \dfrac{4 \cdot 6}{2,598,960} \approx 0.000009$

**51.** (a) $P(\text{no defective}) = \dfrac{_{197}C_3}{_{200}C_3} = \dfrac{1,254,890}{1,313,400} \approx 0.955$

(b) $P(\text{all defective}) = \dfrac{_3C_3}{_{200}C_3} = \dfrac{1}{1,313,400} \approx 0.0000008$

(c) $P(\text{at least one defective}) = 1 - P(\text{no defective}) \approx 1 - 0.955 = 0.045$

(d) $P(\text{at least one non-defective}) = 1 - P(\text{all defective}) \approx 1 - 0.0000008 \approx 0.9999992$

**53.** (a) $P(4 \text{ men}) = \dfrac{_6C_4}{_{10}C_4} = \dfrac{15}{210} \approx 0.071$

(b) $P(4 \text{ women}) = \dfrac{_4C_4}{_{10}C_4} = \dfrac{1}{210} \approx 0.005$

(c) $P(2 \text{ men and } 2 \text{ women}) = \dfrac{\left(_6C_2\right) \cdot \left(_4C_2\right)}{_{10}C_4} = \dfrac{15 \cdot 6}{210} \approx 0.429$

(d) $P(1 \text{ man and } 3 \text{ women}) = \dfrac{\left(_6C_1\right) \cdot \left(_4C_3\right)}{_{10}C_4} = \dfrac{6 \cdot 4}{210} \approx 0.114$

## CHAPTER 3 QUIZ SOLUTIONS

**1.** $(9)(10)(10)(10)(10)(5) = 450,000$

**2.** (a) $P(\text{bachelor's degree}) = \dfrac{1716}{3553} \approx 0.483$

(b) $P(\text{bachelor's degree} \mid \text{female}) = \dfrac{982}{2086} \approx 0.471$

(c) $P(\text{bachelor's degree} \mid \text{male}) = \dfrac{734}{1467} \approx 0.500$

(d) $P(\text{associate's degree or bachelor's degree}) = P(\text{associate's degree}) + P(\text{bachelor's degree})$

$$= \dfrac{942}{3553} + \dfrac{1716}{3553} \approx 0.748$$

(e) $P(\text{doctorate} \mid \text{female}) = \dfrac{84}{2086} \approx 0.040$

(f) $P(\text{master's degree or male}) = P(\text{master's degree}) + P(\text{male}) - P(\text{master's degree and male})$

$$= \dfrac{731}{3553} + \dfrac{1467}{3553} - \dfrac{292}{3553} = \dfrac{1906}{3553} \approx 0.536$$

(g) $P(\text{associate's degree and male}) = \dfrac{361}{3553} \approx 0.102$

(h) $P(\text{female}|\text{bachelor's degree}) = \dfrac{982}{1716} \approx 0.572$

3. The event in part (e) is unusual because its probability is less than or equal to 0.05.

4. Not mutually exclusive. A golfer can score the best round in a four-round tournament and still lose the tournament.
Dependent. The outcome of scoring the best round in a four-round tournament affects the outcome of losing the golf tournament.

5. $_{30}P_4 = \dfrac{30!}{(30-4)!} = 657,720$

6. (a) $_{247}C_3 = \dfrac{247!}{(247-3)!3!} = 2,481,115$

 (b) $_3C_3 = \dfrac{3!}{(3-3)!3!} = 1$

 (c) $\left(_{250}C_3\right) - \left(_3C_3\right) = \dfrac{250!}{(250-3)!3!} - 1 = 2,573,000 - 1 = 2,572,999$

7. (a) $\dfrac{_{247}C_3}{_{250}C_3} = \dfrac{2,481,115}{2,573,000} \approx 0.964$    (b) $\dfrac{_3C_3}{_{250}C_3} = \dfrac{1}{2,573,000} \approx .0000004$

 (c) $\dfrac{_{250}C_3 - {_3}C_3}{_{250}C_3} = \dfrac{2,572,999}{2,573,000} \approx 0.9999996$

# Discrete Probability Distributions

## 4.1 PROBABILITY DISTRIBUTIONS

### 4.1 Try It Yourself Solutions

**1a.** (1) measured      (2) counted

  **b.** (1) The random variable is continuous because $x$ can be any speed up to the maximum speed of a rocket.

     (2) The random variable is discrete because the number of calves born on a farm in one year is countable.

**2ab.**

| $x$ | $f$ | $P(x)$ |
|-----|-----|--------|
| 0 | 16 | 0.16 |
| 1 | 19 | 0.19 |
| 2 | 15 | 0.15 |
| 3 | 21 | 0.21 |
| 4 | 9 | 0.09 |
| 5 | 10 | 0.10 |
| 6 | 8 | 0.08 |
| 7 | 2 | 0.02 |
| | $n = 100$ | $\sum P(x) = 1$ |

**c.**

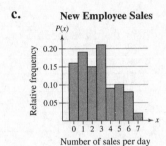

New Employee Sales

**3a.** Each $P(x)$ is between 0 and 1.

  **b.** $\sum P(x) = 1$

  **c.** Because both conditions are met, the distribution is a probability distribution.

**4a.** (1) The probability of each outcome is between 0 and 1.
     (2) The probability of each outcome is between 0 and 1.

  **b.** (1) Yes, $\sum P(x) = 1$.             (2) No, $\sum P(x) = 1.04 \neq 1.00$.

  **c.** (1) Is a probability distribution       (2) Is not a probability distribution.

**5ab.**

| $x$ | $P(x)$ | $xP(x)$ |
|-----|--------|---------|
| 0 | 0.16 | $(0)(0.16) = 0.00$ |
| 1 | 0.19 | $(1)(0.19) = 0.19$ |
| 2 | 0.15 | $(2)(0.15) = 0.30$ |
| 3 | 0.21 | $(3)(0.21) = 0.63$ |
| 4 | 0.09 | $(4)(0.09) = 0.36$ |
| 5 | 0.10 | $(5)(0.10) = 0.50$ |
| 6 | 0.08 | $(6)0.08 = 0.48$ |
| 7 | 0.02 | $(7)(0.02) = 0.14$ |
| | $\sum P(x) = 1$ | $\sum xP(x) = 2.60$ |

  **c.** $\mu = \sum xP(x) = 2.6$

    On average, a new employee makes 2.6 sales per day.

**6ab.** From 5, $\mu = 2.6$,

| $x$ | $f$ | $x - \mu$ | $(x - \mu)^2$ | $P(x)(x - \mu)^2$ |
|-----|-----|-----------|---------------|-------------------|
| 0 | 0.16 | −2.6 | 6.76 | $(0.16)(6.76) = 1.0816$ |
| 1 | 0.19 | −1.6 | 2.56 | $(0.19)(2.56) = 0.4864$ |
| 2 | 0.15 | −0.6 | 0.36 | $(0.15)(0.36) = 0.0540$ |
| 3 | 0.21 | 0.4 | 0.16 | $(0.21)(0.16) = 0.0336$ |
| 4 | 0.09 | 1.4 | 1.96 | $(0.09)(1.96) = 0.1764$ |
| 5 | 0.10 | 2.4 | 5.76 | $(0.10)(5.76) = 0.5760$ |
| 6 | 0.08 | 3.4 | 11.56 | $(0.08)(11.56) = 0.9248$ |
| 7 | 0.02 | 4.4 | 19.36 | $(0.02)(19.36) = 0.3872$ |
| | $\sum P(x) = 1$ | | | $\sum P(x)(x - \mu)^2 = 3.72$ |

$\sigma^2 \approx 3.7$

**c.** $\sigma = \sqrt{\sigma^2} = \sqrt{3.72} \approx 1.9$

**d.** Most of the data values differ from the mean by no more than 1.9 sales per day.

**7ab.**

| Gain, $x$ | $P(x)$ | $xP(x)$ |
|-----------|--------|---------|
| $1995 | $\dfrac{1}{2000}$ | $\dfrac{1995}{2000}$ |
| $ 995 | $\dfrac{1}{2000}$ | $\dfrac{995}{2000}$ |
| $ 495 | $\dfrac{1}{2000}$ | $\dfrac{495}{2000}$ |
| $ 245 | $\dfrac{1}{2000}$ | $\dfrac{245}{2000}$ |
| $ 95 | $\dfrac{1}{2000}$ | $\dfrac{95}{2000}$ |
| $ −5 | $\dfrac{1995}{2000}$ | $-\dfrac{9975}{2000}$ |
| | $\sum P(x) = 1$ | $\sum xP(x) \approx -3.08$ |

**c.** $E(x) = \sum xP(x) = -\$3.08$

**d.** Because the expected value is negative, you can expect to lose an average of $3.08 for each ticket you buy.

## 4.1 EXERCISE SOLUTIONS

1. A random variable represents a numerical value associated with each outcome of a probability experiment. Examples: Answers will vary.

3. No; The expected value may not be a possible value of $x$ for one trial, but it represents the average value of $x$ over a large number of trials.

5. False. In most applications, discrete random variables represent counted data, while continuous random variables represent measured data.

7. False. The mean of the random variable of a probability distribution describes a typical outcome. The variance and standard deviation of the random variable of a probability distribution describe how the outcomes vary.

9. Discrete; Attendance is a random variable that is countable.

11. Continuous; The distance a baseball travels after being hit is a random variable that must be measured.

13. Discrete; The number of books in a university library is a random variable that is countable.

15. Continuous; The volume of blood drawn for a blood test is a random variable that must be measured.

17. Discrete; The number of messages posted each month on a social networking site is a random variable that is countable.

19. (a)

| $x$ | $f$ | $P(x)$ |
|---|---|---|
| 0 | 26 | 0.01 |
| 1 | 442 | 0.17 |
| 2 | 728 | 0.28 |
| 3 | 1404 | 0.54 |
| | $n = 2600$ | $\sum P(x) = 1$ |

(b)

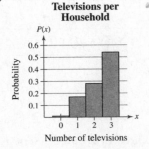

**Televisions per Household**

Skewed left

21. (a) $P(1 \text{ or } 2) = 0.17 + 0.28 = 0.45$

(b) $P(2 \text{ or more}) = 0.28 + 0.54 = 0.82$

(c) $P(1 \leq x \leq 3) = 0.17 + 0.28 + 0.54 = 0.99$

23. $P(0) = 0.01$. Yes, because the probability is less than 0.05.

25. $\sum P(x) = 1 \rightarrow P(3) = 1 - (0.07 + 0.20 + 0.38 + 0.13) = 0.22$

27. Because each $P(x)$ is between 0 and 1, and $\sum P(x) = 1$, the distribution is a probability distribution.

**29.** (a)

| x | P(x) | xP(x) | (x − μ) | (x − μ)² | (x − μ)²P(x) |
|---|------|-------|---------|----------|--------------|
| 0 | 0.686 | 0 | −0.497 | 0.247 | 0.169 |
| 1 | 0.195 | 0.195 | 0.503 | 0.253 | 0.049 |
| 2 | 0.077 | 0.154 | 1.503 | 2.259 | 0.174 |
| 3 | 0.022 | 0.066 | 2.503 | 6.265 | 0.138 |
| 4 | 0.013 | 0.052 | 3.503 | 12.271 | 0.160 |
| 5 | 0.006 | 0.030 | 4.503 | 20.277 | 0.122 |
| | $\sum P(x) \approx 1$ | $\sum xP(x) = 0.497$ | | | $\sum (x - \mu)^2 P(x) = 0.812$ |

$$\mu = \sum xP(x) = 0.497 \approx 0.5$$

$$\sigma^2 = \sum (x - \mu)^2 P(x) = 0.812 \approx 0.8$$

$$\sigma = \sqrt{\sigma^2} = \sqrt{0.812} \approx 0.9$$

(b) The mean is 0.5, so the average number of dogs per household is about 0 or 1 dog. The standard deviation is 0.9, so most of the households differ from the mean by no more than 1 dog.

**31.** (a)

| x | P(x) | xP(x) | (x − μ) | (x − μ)² | (x − μ)²P(x) |
|---|------|-------|---------|----------|--------------|
| 0 | 0.250 | 0 | −1.501 | 2.253 | 0.563 |
| 1 | 0.298 | 0.298 | −0.501 | 0.251 | 0.075 |
| 2 | 0.229 | 0.458 | 0.499 | 0.249 | 0.057 |
| 3 | 0.168 | 0.504 | 1.4995 | 2.247 | 0.377 |
| 4 | 0.034 | 0.136 | 2.499 | 6.245 | 0.212 |
| 5 | 0.021 | 0.105 | 3.499 | 12.243 | 0.257 |
| | | $\sum xP(x) = 1.501$ | | | $\sum (x - \mu)^2 P(x) = 1.542$ |

$$\mu = \sum xP(x) \approx 1.5$$

$$\sigma^2 = \sum (x - \mu)^2 P(x) = 1.542 \approx 1.5$$

$$\sigma = \sqrt{\sigma^2} = \sqrt{1.542} \approx 1.2$$

(b) The mean is 1.5, so the average batch of camping chairs has about 1 or 2 defects. The standard deviation is 1.2, so most of the batches differ from the mean by no more than about 1 defect.

**33.** (a)

| x | P(x) | xP(x) | (x − μ) | (x − μ)² | (x − μ)²P(x) |
|---|------|-------|---------|----------|--------------|
| 1 | 0.418 | 0.418 | −0.983 | 0.966 | 0.404 |
| 2 | 0.261 | 0.522 | 0.017 | 0.000 | 0.000 |
| 3 | 0.247 | 0.807 | 1.017 | 1.034 | 0.255 |
| 4 | 0.063 | 0.256 | 2.017 | 4.068 | 0.256 |
| 5 | 0.010 | 0.055 | 3.017 | 9.102 | 0.091 |
| | | $\sum xP(x) = 1.983$ | | | $\sum (x - \mu)^2 P(x) = 1.007$ |

$$\mu = \sum xP(x) = 1.983 \approx 2.0$$

$$\sigma^2 = \sum (x - \mu)^2 P(x) = 1.007 \approx 1.0$$

$$\sigma = \sqrt{\sigma^2} = \sqrt{1.007} \approx 1.0$$

(b) The mean is 2.0, so an average hurricane that hits the U.S. mainland is expected to be a category 2 hurricane. The standard deviation is 1.0, so most of the hurricanes differ from the mean by no more than 1 category level.

**35.** An expected value of 0 means that the money gained is equal to the money spent, representing the breakeven point.

**37.** $E(x) = \mu = \sum xP(x) = (-\$1) \cdot \left(\dfrac{37}{38}\right) + (\$35) \cdot \left(\dfrac{1}{38}\right) \approx -\$0.05$

**39.** $\mu_x = a + b\mu_x = 1000 + 1.05(36,000) = \$38,800$

**41.** $\mu_{x+y} = \mu_x + \mu_y = 1512 + 1486 = 2998$

$\mu_{x-y} = \mu_x - \mu_y = 1512 - 1486 = 26$

## 4.2 BINOMIAL DISTRIBUTIONS

## 4.2 Try It Yourself Solutions

**1a.** Trial: answering a question
Success: the question answered correctly
**b.** Yes, the experiment satisfies the four conditions of a binomial experiment.
**c.** It is a binomial experiment.
$n = 10,\ p = 0.25,\ q = 0.75,\ x = 0,1,2,3,4,5,6,7,8,9,10$

**2a.** Trial: drawing a card with replacement
Success: card drawn is a club
Failure: card drawn is not a club
**b.** $n = 5,\ p = 0.25,\ q = 0.75,\ x = 3$
**c.** $P(3) = \dfrac{5!}{(5-3)!3!}(0.25)^3(0.75)^2 \approx 0.088$

**3a.** Trial: Selecting an adult and asking a question
Success: selecting an adult who uses a tablet to access social media
Failure: selecting an adult who does not use a tablet to access social media
**b.** $n = 7,\ p = 0.16,\ q = 0.84,\ x = 0,1,2,3,4,5,6,7$
**c.** $P(0) =_7 C_0(0.16)^0(0.84)^7 \approx 0.295090$

$P(1) =_7 C_1(0.16)^1(0.84)^6 \approx 0.393454$

$P(2) =_7 C_2(0.16)^2(0.84)^5 \approx 0.224831$

$P(3) =_7 C_3(0.16)^3(0.84)^4 \approx 0.071375$

$P(4) =_7 C_4(0.16)^4(0.84)^3 \approx 0.013595$

$P(5) =_7 C_5(0.16)^5(0.84)^2 \approx 0.001554$

$P(6) =_7 C_6(0.16)^6(0.84)^1 \approx 0.000099$

$P(7) =_7 C_7 (0.16)^7 (0.84)^0 \approx 0.000003$

**d.**

| $x$ | $P(x)$ |
|-----|--------|
| 0 | 0.295090 |
| 1 | 0.393454 |
| 2 | 0.224831 |
| 3 | 0.071375 |
| 4 | 0.013595 |
| 5 | 0.001554 |
| 6 | 0.000099 |
| 7 | 0.000003 |
| | $\sum xP(x) \approx 1$ |

**4a.** $n = 200$, $p = 0.34$, $x = 68$

**b.** $P(68) \approx 0.059$

**c.** The probability that exactly 68 adults from a random sample of 200 adults with spouses in the United States have hidden purchases from their spouses is about 0.059.

**d.** Because 0.059 is not less than or equal to 0.05, this event is not unusual.

**5a.** (1) $x = 2$    (2) $x = 2, 3, 4,$ or 5    (3) $x = 0$ or 1

**b.** (1) $P(2) =_5 C_2 (0.53)^2 (0.47)^3 \approx 0.292$

(2) $P(3) =_5 C_3 (0.53)^3 (0.47)^2 \approx 0.329$; $P(4) =_5 C_4 (0.53)^4 (0.47)^1 \approx 0.185$;

$P(5) =_5 C_5 (0.53)^5 (0.47)^0 \approx 0.042$;

$P(x \geq 2) = p(2) + p(3) + p(4) + p(5) = 0.292 + 0.329 + 0.185 + 0.042 = 0.848$

(3) $P(0) =_5 C_0 (0.53)^0 (0.47)^5 \approx 0.023$; $P(1) =_5 C_1 (0.53)^1 (0.47)^4 \approx 0.129$;

$P(x < 2) = P(0) + P(1) \approx 0.023 + 0.129 \approx 0.152$

**c.** (1) The probability that exactly two of the five U.S. men believe that there is a link between playing violent video games and teens exhibiting violent behavior is about 0.292.

(2) The probability that at least two of the five U.S. men believe that there is a link between playing violent video games and teens exhibiting violent behavior is about 0.848.

(3) The probability that fewer than two of the five U.S. men believe that there is a link between playing violent video games and teens exhibiting violent behavior is about 0.152.

**6a.** Trial: Selecting a business and asking if it has a website
Success: Selecting a business with a website
Failure: Selecting a business without a website

**b.** $n = 10$, $p = 0.55$, $x = 4$, $P(0) =_4 C_0 (0.19)^0 (0.81)^4 \approx 0.430$

**c.** $P(4) \approx 0.160$

**d.** The probability of randomly selecting 10 small businesses and finding exactly 4 that have a website is 0.160.

**e.** Because 0.160 is greater than 0.05, this event is not unusual.

**7a.**

$$P(1) = {}_4C_1(0.19)^1(0.81)^3 \approx 0.404$$

$$P(2) = {}_4C_2(0.19)^2(0.81)^2 \approx 0.142$$

$$P(3) = {}_4C_3(0.19)^3(0.81)^1 \approx 0.022$$

$$P(4) = {}_4C_4(0.19)^4(0.81)^0 \approx 0.001$$

**b.**

| x | P(x) |
|---|------|
| 0 | 0.430 |
| 1 | 0.404 |
| 2 | 0.142 |
| 3 | 0.022 |
| 4 | 0.001 |

**c.**

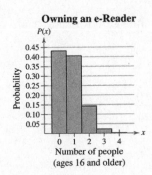

Owning an e-Reader

Skewed right

**d.** Yes, it would be unusual if exactly three or exactly four of the four people owned an e-reader, because each of these events has a probability that is less than 0.05.

**8a.** Success: selecting a clear day
$n = 31, p = 0.44, q = 0.56$

**b.** $\mu = np = (31)(0.44) \approx 13.6$

**c.**

**d.** $\sigma = \sqrt{npq} = \sqrt{(31)(0.44)(0.56)} \approx 2.8 \quad p < 0.5$

**e.** On average, there are about 14 clear days during the month of May. The standard deviation is about 3 days.

**f.** Values that are more than 2 standard deviations from the mean are considered unusual. Because $13.6 - 2(2.8) = 8$ and $13.6 + 2(2.8) = 19.2$, a May with fewer than 8 clear days, or more than 19 clear days would be unusual. $\sigma^2 = npq = (31)(0.44)(0.56) \approx 7.6 \quad p = 0.25$

## 4.2 EXERCISE SOLUTIONS

**1.** Each trial is independent of the other trials when the outcome of one trial does not affect the outcome of any of the other trials.

**3.** (a) $p = 0.75$ (graph is skewed left $\rightarrow p > 0.5$)

   (b) $p = 0.50$ (graph is symmetric)

(c) $p = 0.25$ (graph is skewed right $\rightarrow p < 0.5$ )

As the probability increases, the graph moves from skewed right towards skewed left because a greater probability means that it is more likely for more trials to be successful.

5. (a) $x = 0,1$      (b) $x = 0,5$      (c) $x = 4,5$

7. $\mu = np = (50)(0.4) = 20$

   $\sigma^2 = npq = (50)(0.4)(0.6) = 12$

   $\sigma = \sqrt{npq} = \sqrt{(50)(0.4)(0.6)} \approx 3.5$

9. $\mu = np = (124)(0.26) \approx 32.2$

   $\sigma^2 = npq = (124)(0.26)(0.74) \approx 23.9$

   $\sigma = \sqrt{npq} = \sqrt{(124)(0.26)(0.74)} \approx 4.9$

11. It is a binomial experiment.
    Success: household owns a dedicated game console
    $n = 8$, $p = 0.49$, $q = 0.51$, $x = 0,1,2,3,4,5,6,7,8$

13. Not a binomial experiment because the probability of a success is not the same for each trial.

15. $n = 9$, $p = 0.60$
    (a) $P(5) \approx 0.251$
    (b) $P(x \geq 6) = P(6) + P(7) + P(8) + P(9) \approx 0.251 + 0.161 + 0.060 + 0.010 \approx 0.483$
    (c) $P(x < 4) = P(0) + P(1) + P(2) + P(3) \approx 0.000 + 0.004 + 0.021 + 0.074 \approx 0.099$

17. $n = 12$, $p = 0.27$
    (a) $P(3) \approx 0.255$
    (b) $P(x \geq 4) = P(4) + P(5) + \cdots + P(12)$
    $\approx 0.212 + 0.126 + 0.054 + 0.017 + 0.004 + 0.001 + 0.000 + 0.000 + 0.000 \approx 0.414$
    (c) $P(x < 8) = P(0) + P(1) + \cdots + P(7)$
    $\approx 0.023 + 0.102 + 0.207 + 0.255 + 0.212 + 0.126 + 0.054 + 0.017 \approx 0.995$

19. $n = 8$, $p = 0.56$
    (a) $P(5) \approx 0.263$
    (b) $P(x > 5) = P(6) + P(7) + P(8) \approx 0.167 + 0.061 + 0.010 \approx 0.238$
    (c) $P(x \leq 5) = P(0) + P(1) + P(2) + P(3) + P(4) + P(5)$
    $\approx 0.001 + 0.014 + 0.064 + 0.162 + 0.258 + 0.263 \approx 0.762$

21. $n = 10$, $p = 0.51$
    (a) $P(2) \approx 0.039$
    (b) $P(x > 2) = 1 - P(0) - P(1) - P(2) \approx 1 - 0.001 - 0.008 - 0.039 \approx 0.952$
    (c) $P(2 \leq x \leq 5) = P(2) + P(3) + P(4) + P(5) \approx 0.039 + 0.108 + 0.197 + 0.246 \approx 0.589$

**23.** (a) $n = 7$, $p = 0.67$

| $x$ | $P(x)$ |
|-----|----------|
| 0 | 0.000426 |
| 1 | 0.006057 |
| 2 | 0.036893 |
| 3 | 0.124838 |
| 4 | 0.253460 |
| 5 | 0.308760 |
| 6 | 0.208959 |
| 7 | 0.060607 |

(b) Skewed left

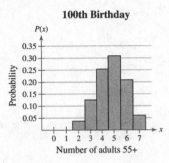

100th Birthday

(c) The values 0, 1, and 2 are unusual because their probabilities are less than 0.05.

**25.** (a) $n = 8$, $p = 0.46$

| $x$ | $P(x)$ |
|-----|----------|
| 0 | 0.007230 |
| 1 | 0.049272 |
| 2 | 0.146905 |
| 3 | 0.250282 |
| 4 | 0.266504 |
| 5 | 0.181618 |
| 6 | 0.077356 |
| 7 | 0.018827 |
| 8 | 0.002005 |

(b) Approximately symmetric

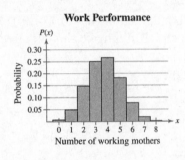

Work Performance

(c) The values 0, 1, 7 and 8 are unusual because their probabilities are less than 0.05.

**27.** (a) $\mu = np = (7)(0.59) \approx 4.1$

(b) $\sigma^2 = npq = (7)(0.59)(0.41) \approx 1.7$

(c) $\sigma = \sqrt{npq} = \sqrt{(7)(0.59)(0.41)} \approx 1.3$

(d) On average, 4.1 out of every 7 U.S. voters think that most school textbooks put political correctness ahead of accuracy. The standard deviation is about 1.3, so most samples of 7 U.S. voters would differ from the mean by at most 1.3 U.S. voters.

**29.** (a) $\mu = np = (6)(0.31) \approx 1.9$

(b) $\sigma^2 = npq = (6)(0.31)(0.69) \approx 1.3$

(c) $\sigma = \sqrt{npq} = \sqrt{(6)(0.31)(0.69)} \approx 1.1$

(d) On average, 1.9 out of every 6 adults think that life existed on Mars at some point in time. The standard deviation is about 1.1, so most samples of 6 adults would differ from the mean by at most 1.1 adults.

**31.** (a) $\mu = np = (6)(0.79) \approx 4.7$

(b) $\sigma^2 = npq = (7)(0.79)(0.21) \approx 1.0$

(c) $\sigma = \sqrt{npq} = \sqrt{(7)(0.79)(0.21)} \approx 1.0$

(d) On average, 4.7 out of every 6 workers know what their CEO looks like. The standard deviation is about 1.0, so most samples of 6 workers would differ from the mean by at most 1.0 worker.

**33.** $P(5,2,2,1) = \dfrac{10!}{5!2!2!1!}\left(\dfrac{9}{16}\right)^5\left(\dfrac{3}{16}\right)^2\left(\dfrac{3}{16}\right)^2\left(\dfrac{1}{16}\right)^1 = 0.033$

## 4.3 MORE DISCRETE PROBABILITY DISTRIBUTIONS

### 4.3 Try It Yourself Solutions

**1a.** $P(1) = (0.75)(0.25)^{1-1} = 0.75$

$P(2) = (0.75)(0.25)^{2-1} \approx 0.188$

**b.** $P(\text{shot made before third attempt}) \approx 0.75 + 0.188 \approx .938$

**c.** The probability that LeBron makes his first free throw shot before his third attempt is 0.938.

**2a.** $P(0) \approx \dfrac{3^0(2.71828)^{-3}}{0!} \approx 0.050$        $P(1) \approx \dfrac{3^1(2.71828)^{-3}}{1!} \approx 0.149$

$P(2) \approx \dfrac{3^2(2.71828)^{-3}}{2!} \approx 0.224$        $P(3) \approx \dfrac{3^3(2.71828)^{-3}}{3!} \approx 0.224$

$P(4) \approx \dfrac{3^4(2.71828)^{-3}}{4!} \approx 0.168$

**b.** $P(0) + P(1) + P(2) + P(3) + P(4) \approx 0.050 + 0.149 + 0.224 + 0.224 + 0.168 = 0.815$

**c.** $1 - 0.815 = 0.185$

**d.** The probability that more than four accidents will occur in any given month at the intersection is 0.185.

**3a.** $\mu = \dfrac{2000}{20,000} = 0.10$

**b.** $\mu = 0.10, x = 3$

**c.** $P(3) = 0.0002$

**d.** The probability of finding three brown trout in any given cubic meter of the lake is 0.0002.

**e.** Because 0.0002 is less than 0.05, this can be considered an unusual event.

### 4.3 EXERCISE SOLUTIONS

**1.** $P(3) = (0.65)(0.35)^{3-1} \approx 0.080$

**3.** $P(5) = (0.09)(0.91)^{5-1} \approx 0.062$

**5.** $P(4) \approx \dfrac{(5)^4(2.71828)^{-5}}{4!} \approx 0.175$

7. $P(2) \approx \dfrac{(1.5)^2 (2.71828)^{-1.5}}{2!} \approx 0.251$

9. In a binomial distribution, the value of $x$ represents the number of success in $n$ trials. In a geometric distribution the value of $x$ represents the first trial that results in a success.

11. $p = 0.19$

   (a) $P(5) = (0.19)(0.81)^{5-1} \approx 0.082$

   (b) $P(\text{sale on } 1^{st}, 2^{nd}, \text{or } 3^{rd} \text{ call}) = P(1) + P(2) + P(3)$
   $$= (0.19)(0.81)^{1-1} + (0.19)(0.81)^{2-1} + (0.19)(0.81)^{3-1} \approx 0.469$$

   (c) $P(\text{no sale on first 3 calls}) = 1 - P(\text{sale on } 1^{st}, 2^{nd}, \text{or } 3^{rd} \text{ call}) \approx 1 - 0.469 \approx 0.531$

13. $\mu = 8$

   (a) $P(5) \approx \dfrac{(8)^5 (2.71828)^{-8}}{5!} \approx 0.092$

   (b) $P(x \geq 5) = 1 - \big(P(0) + P(1) + P(2) + P(3) + P(4)\big)$
   $$\approx 1 - \big(0.000 + 0.003 + 0.011 + 0.029 + 0.057\big) \approx 0.900$$

   (c) $P(x > 5) = 1 - P(x \leq 5) = 1 - \big(P(0) + P(1) + P(2) + P(3) + P(4) + P(5)\big)$
   $$\approx 1 - \big(0.000 + 0.003 + 0.011 + 0.029 + 0.057 + 0.092\big) \approx 0.809$$

15. $p = 0.637$

   (a) $P(2) = (0.637)(0.363)^{2-1} \approx 0.231$.

   (b) $P(\text{completes } 1^{st} \text{ or } 2^{nd} \text{ pass}) = P(1) + P(2)$
   $$= (0.637)(0.363)^{1-1} + (0.637)(0.363)^{2-1} \approx 0.637 + 0.231 \approx 0.868$$

   (c) $P(\text{does not complete his } 1^{st} \text{ two passes}) = 1 - P(\text{completes } 1^{st} \text{ or } 2^{nd} \text{ pass}) \approx 1 - 0.868 = 0.132$

17. $p = \dfrac{1}{500} = 0.002$

   (a) $P(10) = (0.002)(0.998)^{10-1} \approx 0.002$
   This event is unusual because its probability is less than 0.05.

   (b) $P\big(1^{st}, 2^{nd}, \text{ or } 3^{rd} \text{ part is warped}\big) = P(1) + P(2) + P(3)$
   $$= (0.002)(0.998)^{1-1} + (0.002)(0.998)^{2-1} + (0.002)(0.998)^{3-1} \approx 0.002 + 0.002 + 0.002 \approx 0.006$$
   This event is unusual because its probability is less than 0.05.

   (c) $P(x > 10) = 1 - P(x \leq 10) = 1 - \big(P(1) + P(2) + \cdots + P(10)\big)$
   $$\approx 1 - \big(0.002 + 0.002 + 0.002 + 0.002 + 0.002 + 0.002 + 0.002 + 0.002 + 0.002 + 0.002\big)$$
   $$\approx 1 - 0.020 \approx 0.980$$

19. $\mu = 0.6$

   (a) $P(1) \approx \dfrac{(0.6)^1 (2.71828)^{-0.6}}{1!} \approx 0.329$

(b) $P(x \le 1) = P(0) + P(1) \approx 0.549 + 0.329 \approx 0.878$

(c) $P(x > 1) = 1 - P(x \le 1) \approx 1 - 0.878 \approx 0.122$

**21.** $\mu = 6$

(a) $P(7) \approx \dfrac{(6)^7 (2.71828)^{-6}}{7!} \approx 0.138$

(b) $P(x \ge 8) = 1 - P(x \le 7) = 1 - \big(P(0) + P(1) + P(2) + P(3) + P(4) + P(5) + P(6) + P(7)\big)$

$\approx 1 - \big(0.002 + 0.015 + 0.045 + 0.089 + 0.134 + 0.161 + 0.161 + 0.138\big) \approx 1 - 0.744 \approx .256$

(c) $P(x \le 4) = P(0) + P(1) + P(2) + P(3) + P(4) \approx 0.002 + 0.015 + 0.045 + 0.089 + 0.134 \approx 0.285$

**23.** $n = 5$, $p = 0.54$

(a) $P(3) = \dfrac{5!}{2!3!}(0.54)^3(0.46)^2 \approx 0.333$

(b) $P(x < 4) = P(x \le 3) = P(0) + P(1) + P(2) + P(3) \approx 0.021 + 0.121 + 0.284 + 0.333 \approx 0.759$

(c) $P(x \ge 3) = P(3) + P(4) + P(5) \approx 0.333 + 0.196 + 0.046 \approx 0.575$

**25.** $n = 6$, $p = 0.42$

(a) $P(4) = \dfrac{6!}{2!4!}(0.42)^4(0.58)^2 \approx 0.157$

(b) $P(x > 2) = 1 - P(x \le 2) = 1 - \big(P(0) + P(1) + P(2)\big) \approx 1 - \big(0.038 + 0.165 + 0.299\big) \approx 0.497$

(c) $P(x \le 5) = 1 - P(6) \approx 1 - 0.005 \approx 0.995$

**27.** (a) $n = 6000$, $p = \dfrac{1}{2500} = 0.0004$

$P(4) = \dfrac{6000!}{5996!4!}(0.0004)^4(0.9996)^{5996} \approx 0.12542$

(b) $\mu = \dfrac{6000}{2500} = 2.4$ cars with defects per 6000.

$P(4) = \dfrac{(2.4)^4(2.71828)^{-2.4}}{4!} \approx 0.1254$

The results are approximately the same.

**29.** $p = \dfrac{1}{1000} = 0.001$

(a) $\mu = \dfrac{1}{p} = \dfrac{1}{0.001} = 1000$

$\sigma^2 = \dfrac{q}{p^2} = \dfrac{0.999}{(0.001)^2} = 999{,}000$

$\sigma = \sqrt{\sigma^2} \approx 999.5$

(b) 1000 times, because it is the mean.

You would expect to lose money, because, on average, you would win $500 every 1000 times you play the lottery and pay $1000 to play it. So, the net gain would be −$500.

**31.** $\mu = 3.9$

(a) $\sigma^2 = \mu = 3.9$

$\sigma = \sqrt{\sigma^2} \approx 2.0$

The standard deviation is 2.0 strokes, so most of Phil's scores per hole differ from the mean by no more than 2.0 strokes.

(b) For 18 holes, Phil's average would be $(18)(3.9) = 70.2$ strokes.

So, $\mu = 70.2$.

$P(X > 72) = 1 - P(x \leq 72) \approx 1 - 0.615 \approx 0.385$

## CHAPTER 4 REVIEW EXERCISE SOLUTIONS

**1.** Discrete.

**3.** (a)

| $x$ | $f$ | $P(x)$ |
|-----|-----|--------|
| 0 | 30 | 0.189 |
| 1 | 65 | 0.409 |
| 2 | 45 | 0.283 |
| 3 | 15 | 0.094 |
| 4 | 4 | 0.025 |
| | $n = 159$ | $\sum P(x) = 1$ |

(b)

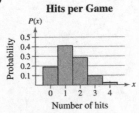

**Hits per Game**

Skewed right

**5.** Yes.

**7.** (a)

| $x$ | $P(x)$ | $xP(x)$ | $(x-\mu)$ | $(x-\mu)^2$ | $(x-\mu)^2 P(x)$ |
|---|---|---|---|---|---|
| 0 | 0.020 | 0 | −2.804 | 7.862 | 0.157 |
| 1 | 0.140 | 0.140 | −1.804 | 3.254 | 0.456 |
| 2 | 0.272 | 0.544 | −0.804 | 0.646 | 0.176 |
| 3 | 0.292 | 0.876 | 0.196 | 0.038 | 0.011 |
| 4 | 0.168 | 0.672 | 1.196 | 1.430 | 0.240 |
| 5 | 0.076 | 0.380 | 2.196 | 4.822 | 0.367 |
| 6 | 0.032 | 0.192 | 3.196 | 10.214 | 0.327 |
| | | $\sum xP(x) = 2.804$ | | | $\sum (x-\mu)^2 P(x) = 1.734$ |

$$\mu = \sum xP(x) = 2.804 \approx 2.8$$

$$\sigma^2 = \sum (x-\mu)^2 P(x) = 1.734 \approx 1.7 \; ; \qquad \sigma = \sqrt{\sigma^2} = \sqrt{1.734} \approx 1.3$$

(b) The mean is 2.8, so the average number of cellular phones per household is about 3. The standard deviation is 1.3, so most of the households differ from the mean by no more than about 1 cellular phone.

**9.**

| $x$ | $P(x)$ | $xP(x)$ |
|---|---|---|
| 100 | 0.125 | 12.500 |
| 0 | 0.250 | 0.000 |
| -25 | 0.625 | -15.625 |
| | | $\sum xP(x) = -3.125$ |

$$\mu = \sum xP(x) = -3.125 \approx -\$3.13$$

**11.** It is a binomial experiment.
Success: a blue candy is selected
$n = 12$, $p = 0.24$, $q = 0.76$, $x = 0,1,2,3,4,5,6,7,8,9,10,11,12$

**13.** $n = 8$, $p = 0.30$

(a) $P(3) \approx 0.254$

(b) $P(x \geq 3) = 1 - P(x < 3) = 1 - (P(0) + P(1) + P(2))$

$\approx 1 - (0.058 + 0.198 + 0.296) \approx 1 - 0.552 \approx 0.448$

(c) $P(x > 3) = 1 - P(x \leq 3) = 1 - (P(0) + P(1) + P(2) + P(3))$

$\approx 1 - (0.058 + 0.198 + 0.296 + 0.254) \approx 1 - 0.806 \approx 0.194$

**15.** $n = 9$, $p = 0.43$

(a) $P(5) \approx 0.196$

(b) $P(x \geq 5) = 1 - P(x < 5) = 1 - (P(0) + P(1) + P(2) + P(3) + P(4))$

$\approx 1 - (0.006 + 0.043 + 0.130 + 0.229 + 0.259) \approx 1 - 0.668 \approx 0.332$

(c) $P(x > 5) = 1 - P(x \leq 5) = 1 - (P(0) + P(1) + P(2) + P(3) + P(4) + P(5))$

$\approx 1 - (0.006 + 0.043 + 0.130 + 0.229 + 0.259 + 0.196) \approx 1 - 0.863 \approx 0.137$

**17.** (a) $n = 5$, $p = 0.38$

| $x$ | $P(x)$ |
|---|---|
| 0 | 0.092 |
| 1 | 0.281 |
| 2 | 0.344 |
| 3 | 0.211 |
| 4 | 0.065 |
| 5 | 0.008 |

(b)

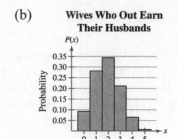

**Wives Who Out Earn Their Husbands**

Skewed right

(c) The value 5 is unusual because its probabilities are less than 0.05.

**19.** $n = 8$, $p = 0.14$

(a) $\mu = np = (8)(0.14) \approx 1.1$

(b) $\sigma^2 = npq = (8)(0.14)(0.86) \approx 1.0$

(c) $\sigma = \sqrt{npq} = \sqrt{(8)(0.14)(0.86)} \approx 1.0$

(d) On average, 1.1 out of every 8 drivers are uninsured. The standard deviation is 1.0, so most samples of 8 drivers would differ from the mean by at most 1 driver.

**21.** $p = 0.22$

(a) $P(3) = (0.22)(0.78)^2 \approx 0.134$

(b) $P(4 \text{ or } 5) = P(4) + P(5) = (0.22)(0.78)^3 + (0.22)(0.78)^4 \approx 0.186$

(c) $P(x > 7) = 1 - P(x \le 7) = 1 - \big(P(1) + P(2) + P(3) + P(4) + P(5) + P(6) + P(7)\big)$

$\approx 1 - \big(0.220 + 0.172 + 0.134 + 0.104 + 0.081 + 0.064 + 0.050\big) \approx 0.175$

**23.** $n = 7$, $p = 0.37$

(a) $P(4) \approx 0.164$

(b) $P(x < 2) = P(0) + P(1) \approx 0.039 + 0.162 \approx 0.201$

(c) $P(x \ge 5) = P(6) + P(7) \approx 0.011 + 0.001 \approx 0.012$

This event would be unusual because its probability is less than 0.05.

**25.** $\mu = 5$

(a) $P(3) \approx \dfrac{(5)^3 (2.71828)^{-5}}{3!} \approx 0.140$

(b) $P(x > 6) = 1 - P(x \le 6) = 1 - \big(P(0) + P(1) + P(2) + P(3) + P(4) + P(5) + P(6)\big)$

$\approx 1 - \big(0.007 + 0.034 + 0.084 + 0.140 + 0.175 + 0.175 + 0.146\big) \approx 0.239$

(c) $P(x \le 5) = P(0) + P(1) + P(2) + P(3) + P(4) + P(5)$

$\approx 0.007 + 0.034 + 0.084 + 0.140 + 0.175 + 0.175 \approx 0.615$

## CHAPTER 4 QUIZ SOLUTIONS

1. (a) Discrete; The number of lightning strikes that occur in Wyoming during the month of June is a random variable that is countable.
   (b) Continuous; The fuel (in gallons) used by a jet during takeoff is a random variable that has an infinite number of possible outcomes and cannot be counted.
   (c) Discrete; The number of die rolls required for an individual to roll a five is a random variable that is countable.

2. (a)

| $x$ | $f$ | $P(x)$ |
|-----|-----|--------|
| 0 | 27 | 0.237 |
| 1 | 47 | 0.412 |
| 2 | 24 | 0.211 |
| 3 | 10 | 0.088 |
| 4 | 4 | 0.035 |
| 5 | 2 | 0.018 |
| | $n = 114$ | $\sum P(x) \approx 1$ |

(b)

**Computers per Household**

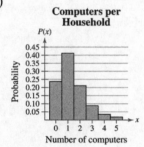

Skewed right

(c)

| $xP(x)$ | $(x - \mu)$ | $(x - \mu)^2$ | $(x - \mu)^2 P(x)$ |
|---------|-------------|---------------|---------------------|
| 0.000 | -1.328 | 1.764 | 0.418 |
| 0.412 | -0.328 | 0.108 | 0.044 |
| 0.422 | 0.672 | 0.452 | 0.095 |
| 0.264 | 1.672 | 2.796 | 0.246 |
| 0.140 | 2.672 | 7.140 | 0.250 |
| 0.090 | 3.672 | 13.484 | 0.243 |
| $\sum xP(x) = 1.328$ | | | $\sum (x - \mu)^2 P(x) = 1.296$ |

$\mu = \sum xP(x) \approx 1.328 \approx 1.3$

$\sigma^2 = \sum \leq (x - \mu)^2 P(x) \approx 1.296 \approx 1.3$

$\sigma = \sqrt{\sigma^2} \approx \sqrt{1.296} \approx 1.1$

The mean is 1.3, so the average number of computers per household is 1.3. The standard deviation is 1.10, so most households will differ from the mean by no more than about 1 computer.

(d) $P(x \geq 4) = P(4) + P(5) = \dfrac{4}{114} + \dfrac{2}{114} \approx 0.053$

3. $n = 9$, $p = 0.44$

   (a) $P(3) \approx 0.221$

   (b) $P(x \le 4) = P(0) + P(1) + P(2) + P(3) + P(4) \approx 0.005 + 0.038 + 0.120 + 0.221 + 0.260 \approx 0.645$

   (c) $P(x > 7) = P(8) + P(9) \approx 0.007 + 0.001 \approx 0.008$

4. $n = 6$, $p = 0.85$

   (a)

| $x$ | $P(x)$ |
|-----|--------|
| 0 | 0.00001 |
| 1 | 0.00039 |
| 2 | 0.00549 |
| 3 | 0.04145 |
| 4 | 0.17618 |
| 5 | 0.39933 |
| 6 | 0.37715 |
| | $\sum P(x) = 1$ |

   (b)

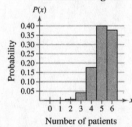

**Successful Surgeries**

Skewed left

   (c) $\mu = np = (6)(0.85) = 5.1$

$$\sigma^2 = npq = (6)(0.85)(0.15) \approx 0.8$$

$$\sigma = \sqrt{npq} = \sqrt{(6)(0.85)(0.15)} \approx 0.9$$

On average, 5.1 out of every 6 patients have a successful surgery. The standard deviation is 0.9, so most samples of 6 surgeries would differ from the mean by at most 0.9 surgery.

5. $\mu = 5$

   (a) $P(5) \approx \dfrac{(5)^5 (2.71828)^{-5}}{5!} \approx 0.175$

   (b) $P(x < 5) = P(0) + P(1) + P(2) + P(3) + P(4) \approx 0.007 + 0.034 + 0.084 + 0.140 + 0.175 \approx 0.440$

   (c) $P(0) \approx \dfrac{(5)^0 (2.71828)^{-5}}{0!} \approx 0.007$

6. $p = 0.58$

   (a) $P(4) \approx (0.58)(0.42)^3 \approx 0.043$

   (b) $P(2^{nd} \text{ or } 3^{rd}) = P(2) + P(3) \approx (0.58)(0.42)^1 + (0.58)(0.42)^2 \approx 0.244 + 0.102 \approx 0.346$

(c) $P(x > 3) = 1 - P(x \le 3) = 1 - \big(P(1) + P(2) + P(3)\big)$

$\approx 1 - \big((0.58)(0.42)^0 + (0.58)(0.42)^1 + (0.58)(0.42)^2\big) \approx 1 - 0.926 \approx 0.074$

7. Event (a) is unusual because its probability is less than 0.05.

## 5.1 INTRODUCTION TO NORMAL DISTRIBUTIONS AND THE STANDARD NORMAL DISTRIBUTION

### 5.1 Try It Yourself Solutions

**1a.** $A$: $x = 45$, $B$: $x = 60$, $C$: $x = 45$ ($B$ has the greatest mean.)
  **b.** Curve $C$ is more spread out, so curve $C$ has the greatest standard deviation.

**2a.** $x = 655$; Thus, $\mu = 45$
  **b.** Inflection points: 635 and 675
   Standard deviation = 20

**3a.** (1) 0.0143          (2) 0.9850

**4a.**

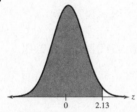

  **b.** 0.9834

**5a.**

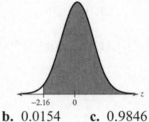

  **b.** 0.0154     **c.** 0.9846

**6a.**

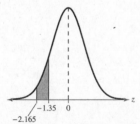

  **b.** 0.0885     **c.** 0.0152
  **c.** Area $= 0.0885 - 0.0152 = 0.0733$

### 5.1 EXERCISE SOLUTIONS

**1.** Answers will vary.

**3.** 1

**5.** Answers will vary.
   Similarities: The two curves will have the same line of symmetry.
   Differences: The curve with the larger standard deviation will be more spread out than the curve with the smaller standard deviation.

**7.** $\mu = 0$, $\sigma = 1$

**9.** "The" standard normal distribution is used to describe one specific normal distribution ($\mu = 0$, $\sigma = 1$). "A" normal distribution is used to describe a normal distribution with any mean and standard deviation.

**11.** No, the graph crosses the $x$-axis.

**13.** Yes, the graph fulfills the properties of the normal distribution. $\mu \approx 18.5$, $\sigma \approx 2$

**15.** No, the graph is skewed to the right.

**17.** (Area left of $z = -1.3$) = 0.0968

**19.** (Area right of $z = 2$) = $1 - $ (Area leftt of $z = 2$) = $1 - 0.9772 = 0.0228$

**21.** (Area left of $z = 0$) $-$ (Area left of $z = -2.25$) = $0.5000 - 0.0122 = 0.4878$

**23.** 0.5319        **25.** 0.0050        **27.** $1 - 0.2578 = 0.7422$

**29.** $1 - 0.3613 = 0.6387$    **31.** $0.9979 - 0.5000 = 0.4979$   **33.** $0.9750 - 0.0250 = 0.9500$

**35.** $0.1003 + 0.1003 = 0.2006$

**37.** (a)

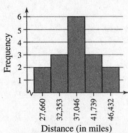

**Life Spans of Tires**

It is reasonable to assume that the life spans are normally distributed because the histogram is symmetric and bell-shaped.
(b) $\bar{x} = 37,234.7$; $s = 6259.2$
(c) The sample mean of 37,234.7 hours is less than the claimed mean, so, on average, the tires in the sample lasted for a shorter time. The sample standard deviation of 6259.2 is greater than the claimed standard deviation, so the tires in the sample had greater variation in life span than the manufacturer's claim.

**39.** (a)   $x = 1920 \Rightarrow z = \dfrac{x - \mu}{\sigma} = \dfrac{1920 - 1498}{316} \approx 1.34$

        $x = 1240 \Rightarrow z = \dfrac{x - \mu}{\sigma} = \dfrac{1240 - 1498}{316} \approx -0.82$

        $x = 2220 \Rightarrow z = \dfrac{x - \mu}{\sigma} = \dfrac{2220 - 1498}{316} \approx 2.22$

$$x = 1390 \Rightarrow z = \frac{x - \mu}{\sigma} = \frac{1390 - 1498}{316} \approx -0.34$$

(b) $x = 2200$ is unusual because its corresponding $z$-score $(2.22)$ lies more than 2 standard deviations from the mean.

**41.** 0.9750　　　　　　**43.** $1 - 0.0225 = 0.9775$　　　　　　**45.** $0.8413 - 0.1587 = 0.6826$

**47.** $P(z < 1.45) = 0.9265$

**49.** Using technology: $P(z > 2.175) = 1 - P(z < 2.175) = 1 - 0.9852 = 0.0148$

**51.** $P(-0.89 < z < 0) = 0.5 - 0.1867 = 0.3133$

**53.** $P(-1.65 < z < 1.65) = 0.9505 - 0.0495 = 0.9010$

**55.** $P(z < -2.58 \text{ or } z > 2.58) = 2(0.0049) = 0.0098$

**57.**

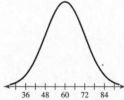

The normal distribution curve is centered at its mean (60) and has 2 points of inflection (48 and 72) representing $\mu \pm \sigma$.

**59.** The area under the curve is $(b - a)\left(\dfrac{1}{b - a}\right) = \dfrac{b - a}{b - a} = 1$.

(Because a < b, you do not have to worry about division by 0.)

## 5.2 NORMAL DISTRIBUTIONS: FINDING PROBABILITIES

### 5.2 Try It Yourself Solutions

**1a.**

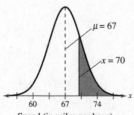

Speed (in miles per hour)

**b.** $z = \dfrac{x - \mu}{\sigma} = \dfrac{70 - 67}{3.5} \approx 0.86$

**c.** $P(z < 0.86) = 0.8051$

$P(z > 0.86) = 1 - P(z < 0.86) \approx 1 - 0.8051 \approx 0.1949$

**d.** The probability that a randomly selected vehicle is violating the speed limit of 70 miles per hour is 0.1949.

**2a.**

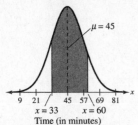

$x = 33 \qquad x = 60$

Time (in minutes)

**b.** $z = \dfrac{x - \mu}{\sigma} = \dfrac{33 - 45}{12} = -1$

$z = \dfrac{x - \mu}{\sigma} = \dfrac{60 - 45}{12} = 1.25$

**c.** $P(z < -1) = 0.1587$

$P(z < 1.25) = 0.8944$

$0.8944 - 0.1587 = 0.7357$

**d.** When 150 shoppers enter the store, you would expect $150(0.7357) = 110.355$ or about 110 shoppers to be in the store between 33 and 60 minutes.

**3a.** Read user's guide for the technology tool.

**b.** Enter the data.

$P(100 < x < 150) = P(-0.97 < z < 0.46) = 0.6762 - 0.1657 = 0.5105$

**c.** The probability that a randomly selected person's triglyceride level is between 100 and 150 is 0.5105.

## 5.2 EXERCISE SOLUTIONS

**1.** $P(x < 170) = P(z < -0.2) = 0.4207$

**3.** $P(x > 182) = P(z > 0.4) = 1 - 0.6554 = 0.3446$

**5.** $P(160 < x < 170) = P(-0.7 < z < -0.2) = 0.4207 - 0.2420 = 0.1787$

**7.** (a) Using technology: $P(x < 66) \approx P(z < -1.1724) = 0.1205$

(b) Using technology: $P(66 < x < 72) = P(-1.1724 < z < 0.8966)$

$= 0.81502 - 0.12052 = 0.6945$

(c) Using technology: $P(x > 72) = P(z > 0.8966) = 1 - P(z < 0.8966) = 1 - 0.8150 = 0.1850$

(d) No, none of these events are unusual because their probabilities are greater than 0.05.

**9.** (a) Using technology: $P(x < 15) = P(z < -1.0161) = 0.1548$

(b) Using technology: $P(18 < x < 25) = P(-0.5323 < z < 0.5968) = 0.72467 - 0.29727 = 0.4274$

(c) Using technology: $P(x > 34) = P(z > 2.0484) = 1 - P(z < 2.0484) = 1 - 0.9797 = 0.0203$

(d) Yes, the event in part (c) is unusual because its probability is less than 0.05.

**11.** (a) Using technology: $P(x < 70) = P(z < -2.5) = 0.0062$

(b) Using technology: $P(90 < x < 120) = P(-0.8333 < z < 1.6667)$
$$= 0.95221 - 0.20233 = 0.7499$$

(c) Using technology: $P(x > 140) = P(z > 3.3333) = 1 - P(z < 3.3333) = 1 - 0.9996 = 0.0004$

**13.** Using technology: $P(200 < x < 450) = P(-2.5263 < z < -0.3333)$
$$= 0.36944 - 0.00576 = 0.3637$$

**15.** Using technology: $P(220 < x < 255) = P(0.3968 < z < 1.3228) = 0.90704 - 0.65425 = 0.2528$

**17.** (a) Using technology: $P(x < 600) = P(z < 0.9825) = 0.8371 \Rightarrow 83.71\%$

(b) Using technology: $P(x > 500) = P(z > 0.1053 = 1 - P(z < 0.1053) = 1 - 0.5419 = 0.4581$
$(1000)(0.4581) = 458.1 \Rightarrow 458$ scores

**19.** (a) Using technology: $P(x < 225) = P(z < 0.5291) = 0.7016 \Rightarrow 70.16\%$

(b) Using technology: $P(x > 260) = P(z > 1.4550) = 1 - P(z < 1.4550) = 1 - 0.9272 = 0.0728$
$(250)(0.0728) = 18.2 \Rightarrow 18$ men

**21.** Out of control, because the 10th observation is more than 3 standard deviations beyond the mean.

**23.** Out of control, because there are nine consecutive points below the mean, and two out of three consecutive points more than 2 standard deviations from the mean.

## 5.3 NORMAL DISTRIBUTIONS: FINDING VALUES

### 5.3 Try It Yourself Solutions

**1ab.** (1)

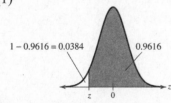

(2)

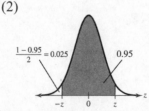

    **c.** (1) $z = -1.77$         (2) $z = \pm 1.96$

**2a.** (1)

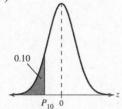

(2)

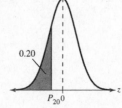

(3)

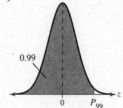

    **b.** (1) use area = 0.1003       (2) use area = 0.2005       (3) use area = 0.9901
    **c.** (1) $z = -1.28$             (2) $z = -0.84$            (3) $z = 2.33$

**3a.** $\mu = 52$, $\sigma = 15$

  **b.** $z = -2.33 \Rightarrow x = \mu + z\sigma = 52 + (-2.33)(15) = 17.05$

      $z = 3.10 \Rightarrow x = \mu + z\sigma = 52 + (3.10)(15) = 98.50$

      $z = 0.58 \Rightarrow x = \mu + z\sigma = 52 + (0.58)(15) = 60.70$

  **c.** 17.05 pounds is below the mean, 60.7 pounds and 98.5 pounds are above the mean.

**4a.**

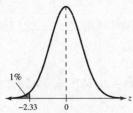

  **b.** $z = -2.33$

  **c.** $x = \mu + z\sigma = 129 + (-2.33)(5.18) \approx 116.93$

  **d.** So, the longest braking distance one of these cars could have and still be in the bottom 1% is about 117 feet.

**5a.**

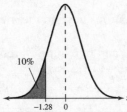

  **b.** $z = -1.28$

  **c.** $x = \mu + z\sigma = 11.2 + (-1.28)(2.1) = 8.512$

  **d.** The maximum length of time an employee could have worked and still be laid off is about 8.5 years.

## 5.3 EXERCISE SOLUTIONS

**1.** $z = -0.81$         **3.** $z = 2.39$         **5.** $z = -1.645$

**7.** $z = 1.555$        **9.** $z = -1.04$        **11.** $z = 1.175$

**13.** $z = -0.67$       **15.** $z = 0.67$        **17.** $z = -0.38$

**19.** $z = -0.58$       **21.** $z = \pm 1.645$

**23.** $\Rightarrow z = -1.18$

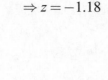

**25.** $\Rightarrow z = 1.18$

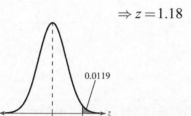

**27.** $\Rightarrow z = \pm 1.28$

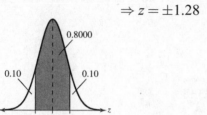

**29.** $\Rightarrow z = \pm 0.06$

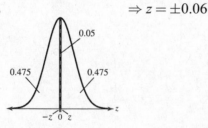

**31.** (a) 95th percentile $\Rightarrow$ Area = 0.95 $\Rightarrow z \approx 1.6449$ (using technology)

      $x = \mu + z\sigma = 64.2 + (1.6449)(2.9) \approx 68.97$ inches

    (b) 1st quartile $\Rightarrow$ Area = 0.25 $\Rightarrow z \approx -0.6745$ (using technology)

      $x = \mu + z\sigma = 64.2 + (-0.6745)(2.9) \approx 62.24$ inches

**33.** (a) 5th percentile $\Rightarrow$ Area = 0.05 $\Rightarrow z \approx -1.6449$ (using technology)

      $x = \mu + z\sigma = 203 + (-1.6449)(25.7) \approx 160.73$ days

    (b) 3rd quartile $\Rightarrow$ Area = 0.75 $\Rightarrow \Rightarrow z \approx 0.6745$ (using technology)

      $x = \mu + z\sigma = 203 + (0.6745)(25.7) \approx 220.33$ days

**35.** (a) Top 5% $\Rightarrow$ Area = 0.95 $\Rightarrow z \approx 1.6449$ (using technology)

      $x = \mu + z\sigma = 6.1 + (1.6449)(1.0) \approx 7.74$ hours

    (b) Middle 50% $\Rightarrow$ Area = 0.25 to 0.75 $\Rightarrow z \approx \pm 0.6745$ (using technology)

      $x = \mu + z\sigma = 6.1 + (\pm 0.6745)(1.0) \approx 5.43$ to $6.77$ hours

**37.** (a) Upper 25% $\Rightarrow$ Area = 0.75 $\Rightarrow$ $z \approx 0.6745$ (using technology)

$x = \mu + z\sigma = 9.5 + (0.6745)(2.8) \approx 11.39$ pounds

(b) Bottom 15% $\Rightarrow$ Area = 0.15 $\Rightarrow$ $z \approx -1.0364$ (using technology)

$x = \mu + z\sigma = 9.5 + (-1.0364)(2.8) \approx 6.60$ pounds

**39.** Upper 4.5% $\Rightarrow$ Area = 0.955 $\Rightarrow$ $z \approx 1.6954$ (using technology)

$x = \mu + z\sigma = 32 + (1.6954)(0.36) \approx 32.61$ ounces

**41.** Top 1% $\Rightarrow$ Area = 0.99 $\Rightarrow$ $z \approx 2.3264$ (using technology)

$x = \mu + z\sigma \Rightarrow 8 = \mu + (2.3264)(0.03) \Rightarrow \mu \approx 7.93$ ounces

## 5.4 SAMPLING DISTRIBUTIONS AND THE CENTRAL LIMIT THEOREM

### 5.4 Try It Yourself Solutions

**1a.**

| Sample | Mean | Sample | Mean | Sample | Mean |
|--------|------|--------|------|--------|------|
| 1, 1, 1 | 1 | 3, 1, 1 | 1.67 | 5, 1, 1 | 2.33 |
| 1, 1, 3 | 1.67 | 3, 1, 3 | 2.33 | 5, 1, 3 | 3 |
| 1, 1, 5 | 2.33 | 3, 1, 5 | 3 | 5, 1, 5 | 3.67 |
| 1, 3, 1 | 1.67 | 3, 3, 1 | 2.33 | 5, 3, 1 | 3 |
| 1, 3, 3 | 2.33 | 3, 3, 3 | 3 | 5, 3, 3 | 3.67 |
| 1, 3, 5 | 3 | 3, 3, 5 | 3.67 | 5, 3, 5 | 4.33 |
| 1, 5, 1 | 2.33 | 3, 5, 1 | 3 | 5, 5, 1 | 3.67 |
| 1, 5, 3 | 3 | 3, 5, 3 | 3.67 | 5, 5, 3 | 4.33 |
| 1, 5, 5 | 3.67 | 3, 5, 5 | 4.33 | 5, 5, 5 | 5 |

**b.**

| $\bar{x}$ | $f$ | Probability |
|------|------|-------------|
| 1 | 1 | 0.03704 |
| 1.67 | 3 | 0.11111 |
| 2.33 | 6 | 0.22222 |
| 3 | 7 | 0.25926 |
| 3.67 | 6 | 0.22222 |
| 4.33 | 3 | 0.11111 |
| 5 | 1 | 0.03704 |
| | $n = 27$ | $\sum P(x) = 1$ |

$\mu_{\bar{x}} = 3$, $\sigma_{\bar{x}}^2 \approx 0.889$, $\sigma_{\bar{x}} \approx 0.943$

**c.** $\mu_{\bar{x}} = \mu = 3,$

$$\sigma_{\bar{x}}^2 = \frac{\sigma^2}{n} = \frac{8/3}{3} \approx 0.889,$$

$$\sigma_{\bar{x}} = \frac{\sigma}{\sqrt{n}} = \frac{\sqrt{8/3}}{\sqrt{3}} \approx 0.943$$

**2a.** $\mu_{\bar{x}} = \mu = 47 \quad \sigma_{\bar{x}} = \frac{\sigma}{\sqrt{n}} = \frac{9}{\sqrt{64}} \approx 1.1$

**b.**

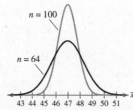

Mean of cell phone bills (in dollars)

**c.** With a smaller sample size, the mean stays the same but the standard deviation increases.

**3a.** $\mu_{\bar{x}} = \mu = 3.5, \quad \sigma_{\bar{x}} = \frac{\sigma}{\sqrt{n}} = \frac{0.2}{\sqrt{16}} = 0.05$

**b.**

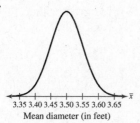

Mean diameter (in feet)

**4a.** $\mu_{\bar{x}} = \mu = 25, \quad \sigma_{\bar{x}} = \frac{\sigma}{\sqrt{n}} = \frac{1.5}{\sqrt{100}} = 0.15$

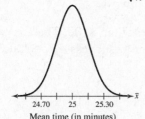

Mean time (in minutes)

**b.** $\bar{x} = 24.7: \; z = \dfrac{\bar{x} - \mu}{\dfrac{\sigma}{\sqrt{n}}} = \dfrac{24.7 - 25}{\dfrac{1.5}{\sqrt{100}}} = -\dfrac{0.3}{0.15} = -2$

$\bar{x} = 25.5: \; z = \dfrac{\bar{x} - \mu}{\dfrac{\sigma}{\sqrt{n}}} = \dfrac{25.5 - 25}{\dfrac{1.5}{\sqrt{100}}} = -\dfrac{0.5}{0.15} \approx 3.33$

  **c.** $P(z < -2) = 0.0228$

  $P(z < 3.33) = 0.9996$

  $P(24.7 < \bar{x} < 25.5) = P(-2 < z < 3.33) = 0.9996 - 0.0228 = 0.9768$

 **d.** Of the samples of 100 drivers ages 15 to 19, about 97.68% will have a mean driving time that is between 24.7 and 25.5 minutes.

**5a.** $\mu_{\bar{x}} = \mu = 176{,}800, \ \sigma_{\bar{x}} = \dfrac{\sigma}{\sqrt{n}} = \dfrac{50{,}000}{\sqrt{12}} \approx 14{,}433.76$

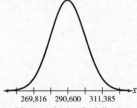

269,816  290,600  311,385

Mean sales price (in dollars)

 **b.** $\bar{x} = 160000: \ z = \dfrac{\bar{x} - \mu}{\dfrac{\sigma}{\sqrt{n}}} = \dfrac{160{,}000 - 176{,}800}{\dfrac{50{,}000}{\sqrt{12}}} \approx \dfrac{-16{,}800}{14{,}433.76} \approx -1.16$

 **c.** $P(z < -1.16) = 0.1230$

  $P(\bar{x} > 160{,}000) = P(z > -1.16) = 1 - P(z < -1.16) = 1 - 0.1230 = 0.8770$

 **d.** About 88% of samples of 12 single-family houses will have a mean sales price greater than $160,000.

**6a.** $x = 200: \ z = \dfrac{x - \mu}{\sigma} = \dfrac{200 - 190}{48} \approx 0.21$

  $\bar{x} = 200: \ z = \dfrac{\bar{x} - \mu}{\dfrac{\sigma}{\sqrt{n}}} = \dfrac{200 - 190}{\dfrac{48}{\sqrt{10}}} \approx \dfrac{10}{15.18} \approx 0.66$

 **b.** $P(z < 0.21) = 0.5832$

  $P(z < 0.66) = 0.7454$

 **c.** There is about a 58% chance that an LCD computer monitor will cost less than $200. There is about a 75% chance that the mean of a sample of 10 LCD computer monitors is less than $200.

## 5.4 EXERCISE SOLUTIONS

**1.** $\mu_{\bar{x}} = \mu = 150$           **3.** $\mu_{\bar{x}} = \mu = 150$           **4.**

$\sigma_{\bar{x}} = \dfrac{\sigma}{\sqrt{n}} = \dfrac{25}{\sqrt{50}} \approx 3.536$      $\sigma_{\bar{x}} = \dfrac{\sigma}{\sqrt{n}} = \dfrac{25}{\sqrt{250}} \approx 1.581$

**5.** False. As the size of a sample increases, the mean of the distribution of sample means does not change.

**7.** False. A sampling distribution is normal if either $n \geq 30$ or the population is normal.

**9.** (c) Because $\mu_{\bar{x}} = 16.5$, $\sigma_{\bar{x}} = \dfrac{\sigma}{\sqrt{n}} = \dfrac{11.9}{\sqrt{100}} = 1.19$, and the graph approximates a normal curve.

**11.**

| Sample | Mean | Sample | Mean | Sample | Mean |
|---|---|---|---|---|---|
| 501, 501 | 501.0 | 575, 501 | 538.0 | 636, 501 | 568.5 |
| 501, 546 | 523.5 | 575, 546 | 560.5 | 636, 546 | 591.0 |
| 501, 575 | 538.0 | 575, 575 | 575.0 | 636, 575 | 605.5 |
| 501, 602 | 551.5 | 575, 602 | 588.5 | 636, 602 | 619.0 |
| 501, 636 | 568.5 | 575, 636 | 605.5 | 636, 636 | 636.0 |
| 546, 501 | 523.5 | 602, 501 | 551.5 | | |
| 546, 546 | 546.0 | 602, 546 | 574.0 | | |
| 546, 575 | 560.5 | 602, 575 | 588.5 | | |
| 546, 602 | 574.0 | 602, 602 | 602.0 | | |
| 546, 636 | 591.0 | 602, 636 | 619.0 | | |

$\mu = 572$, $\sigma \approx 46.31$

$\mu_{\bar{x}} = 572$, $\sigma_{\bar{x}} \approx \dfrac{46.31}{\sqrt{2}} \approx 32.74$

The means are equal but the standard deviation of the sampling distribution is smaller.

**13.**

| Sample | Mean | Sample | Mean | Sample | Mean |
|---|---|---|---|---|---|
| 93, 93, 93 | 93.00 | 95, 93, 93 | 93.67 | 98, 93, 93 | 94.67 |
| 93, 93, 95 | 93.67 | 95, 93, 95 | 94.33 | 98, 93, 95 | 95.33 |
| 93, 93, 98 | 94.67 | 95, 93, 98 | 95.33 | 98, 93, 98 | 96.33 |
| 93, 95, 93 | 93.67 | 95, 95, 93 | 94.33 | 98, 95, 93 | 95.33 |
| 93, 95, 95 | 94.33 | 95, 95, 95 | 95.00 | 98, 95, 95 | 96.00 |
| 93, 95, 98 | 95.33 | 95, 95, 98 | 96.00 | 98, 95, 98 | 97.00 |
| 93, 98, 93 | 94.67 | 95, 98, 93 | 95.33 | 98, 98, 93 | 96.33 |
| 93, 98, 95 | 95.33 | 95, 98, 95 | 96.00 | 98, 98, 95 | 97.00 |
| 93, 98, 98 | 96.33 | 95, 98, 98 | 97.00 | 98, 98, 98 | 98.00 |

$\mu \approx 95.3$, $\sigma \approx 2.05$

$\mu_{\bar{x}} \equiv 95.3$, $\sigma_{\bar{x}} \approx \dfrac{2.05}{\sqrt{3}} \approx 1.18$

The means are equal but the standard deviation of the sampling distribution is smaller.

**15.** $z = \dfrac{\bar{x} - \mu}{\dfrac{\sigma}{\sqrt{n}}} = \dfrac{24.3 - 24}{\dfrac{1.25}{\sqrt{64}}} \approx \dfrac{0.3}{0.156} \approx 1.92$

$P(\bar{x} < 24.3) = P(z < 1.92) = 0.9726$

The probability is not unusual because it is greater than 0.05.

**17.** $z = \dfrac{\bar{x} - \mu}{\dfrac{\sigma}{\sqrt{n}}} = \dfrac{551 - 550}{\dfrac{3.7}{\sqrt{45}}} \approx \dfrac{1}{0.552} \approx 1.8130$

$P(\bar{x} > 551) = P(z > 1.8130) = 1 - P(z < 1.8130) \approx 1 - 0.9651 = 0.0349$ (using technology)

The probability is unusual because it is less than 0.05.

**19.** $\mu_{\bar{x}} = 154$

$\sigma_{\bar{x}} = \dfrac{\sigma}{\sqrt{n}} = \dfrac{5.12}{\sqrt{12}} \approx 1.478$

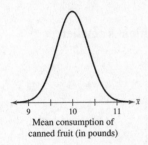

148 150 152 154 156 158 160   $\bar{x}$
Mean braking distance (in feet)

**21.** $\mu_{\bar{x}} = 498$

$\sigma_{\bar{x}} = \dfrac{\sigma}{\sqrt{n}} = \dfrac{116}{\sqrt{20}} \approx 25.938$

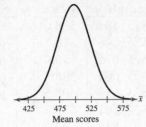

425   475   525   575   $\bar{x}$
Mean scores

**23.** $\mu_{\bar{x}} = 10$

$\sigma_{\bar{x}} = \dfrac{\sigma}{\sqrt{n}} = \dfrac{1.8}{\sqrt{25}} = 0.36$

9    10    11   $\bar{x}$
Mean consumption of
canned fruit (in pounds)

**25.** $n = 24$: $\mu_{\bar{x}} = 154$, $\sigma_{\bar{x}} = \dfrac{\sigma}{\sqrt{n}} = \dfrac{5.12}{\sqrt{24}} \approx 1.045$

$\qquad n = 36$: $\mu_{\bar{x}} = 154$, $\sigma_{\bar{x}} = \dfrac{\sigma}{\sqrt{n}} = \dfrac{5.12}{\sqrt{36}} \approx 0.853$

$n = 24$   $n = 36$

$n = 12$

150   152   154   156   158   $\bar{x}$
Mean braking distance (in feet)

As the sample size increases, the standard deviation of the sample means decreases, while the mean of the sample means remains constant.

**27.** $z = \dfrac{\bar{x} - \mu}{\dfrac{\sigma}{\sqrt{n}}} = \dfrac{60{,}000 - 66{,}000}{\dfrac{12{,}000}{\sqrt{35}}} \approx \dfrac{-6000}{2028.37} \approx -2.96$

$P(\bar{x} < 60{,}000) = P(z < -2.96) = 0.0015$ (using technology)

Only 0.15% of samples of 35 specialists will have a mean salary less than \$60,000. This is an extremely unusual event.

**29.** $z = \dfrac{\bar{x} - \mu}{\dfrac{\sigma}{\sqrt{n}}} = \dfrac{3.781 - 3.796}{\dfrac{0.045}{\sqrt{32}}} \approx \dfrac{-0.015}{0.00795} \approx -1.8856$

$z = \dfrac{\bar{x} - \mu}{\dfrac{\sigma}{\sqrt{n}}} = \dfrac{3.811 - 3.796}{\dfrac{0.045}{\sqrt{32}}} \approx \dfrac{0.015}{0.00795} \approx 1.8856$

$P(3.781 < \bar{x} < 3.811) = P(-1.8856 < z < 1.8856) \approx 0.97033 - 0.02967 = 0.9407$

(using technology)

About 94% of samples of 32 gas stations that week will have a mean price between \$3.781 and \$3.811.

**31.** $z = \dfrac{\bar{x} - \mu}{\dfrac{\sigma}{\sqrt{n}}} = \dfrac{66 - 64.2}{\dfrac{2.9}{\sqrt{60}}} \approx \dfrac{1.8}{0.374} \approx 4.8078$

$P(\bar{x} > 66) = P(z > 4.8078) \approx 0.0000008$ (using technology)

There is almost no chance that a random sample of 60 women will have a mean height greater than 66 inches. This event is almost impossible.

**33.** $z = \dfrac{x - \mu}{\sigma} = \dfrac{70 - 64.2}{2.9} = 2.00$

$P(x < 70) = P(z < 2) = 0.9773$

$z = \dfrac{\bar{x} - \mu}{\dfrac{\sigma}{\sqrt{n}}} = \dfrac{70 - 64.2}{\dfrac{2.9}{\sqrt{20}}} \approx \dfrac{5.8}{0.648} \approx 8.9443$

$P(\bar{x} < 70) = P(z < 8.9443) \approx 1$ (using technology)

It is more likely to select a sample of 20 women with a mean height less than 70 inches because the sample of 20 has a higher probability.

**35.** $z = \dfrac{\bar{x} - \mu}{\dfrac{\sigma}{\sqrt{n}}} = \dfrac{127.9 - 128}{\dfrac{0.20}{\sqrt{40}}} \approx \dfrac{-0.1}{0.0316} \approx -3.1623$

$P(\bar{x} < 127.9) = P(z < -3.1623) \approx 0.0008$ (using technology)

Yes, it is very unlikely that you would have randomly sampled 40 cans with a mean less than or equal to 127.9 ounces because it is more than 3 standard deviations from the mean of the sample means.

**37.** (a) $\mu = 96; \sigma = 0.5$

$z = \dfrac{x - \mu}{\sigma} = \dfrac{96.25 - 96}{0.5} = 0.5$

$P(x > 96.25) = P(z > 0.5) = 1 - P(z < 0.5) \approx 1 - 0.6915 = 0.3085$

(using technology)

(b) $z = \dfrac{\overline{x} - \mu}{\dfrac{\sigma}{\sqrt{n}}} = \dfrac{96.25 - 96}{\dfrac{0.5}{\sqrt{40}}} \approx \dfrac{0.25}{0.079} \approx 3.1623$

$P(\overline{x} > 96.25) = P(z > 3.1623) = 1 - P(z < 3.1623) \approx 1 - 0.9992 = 0.0008$

(using technology)

(c) Although there is about a 31% chance that a board cut by the machine will have a length greater than 96.25 inches, there is less than a 1% chance that the mean of a sample of 40 boards cut by the machine will have a length greater than 96.25 inches. Because there is less than a 1% chance that the mean of a sample of 40 boards will have a length greater than 96.25 inches, this is an unusual event.

**39.** Use the finite correction factor because $n = 55 > 0.05(900) = 45$.

$z = \dfrac{\overline{x} - \mu}{\dfrac{\sigma}{\sqrt{n}} \sqrt{\dfrac{N-n}{N-1}}} = \dfrac{3.742 - 3.746}{\dfrac{0.009}{\sqrt{55}} \sqrt{\dfrac{900 - 55}{900 - 1}}} \approx \dfrac{-0.004}{(0.00121)\sqrt{0.9399}} \approx -3.3998$

$P(\overline{x} < 3.742) = P(z < -3.3998) \approx 0.0003$

(using technology)

**41.** $\mu = p = 0.63;\ \sigma = \sqrt{\dfrac{pq}{n}} = \sqrt{\dfrac{(0.63)(0.37)}{105}} \approx 0.0471$

$z = \dfrac{\hat{p} - p}{\sqrt{\dfrac{pq}{n}}} = \dfrac{0.55 - 0.63}{\sqrt{\dfrac{0.63(0.37)}{105}}} = \dfrac{-0.08}{0.0471} = -1.6979$

$P(\hat{p} < 0.55) = P(z < -1.6979) = 0.0448$

(using technology)

The probability that less than 55% of a sample of 105 residents are in favor of building a new high school is about 4.5%. Because the probability is less than 0.05, this is an unusual event.

## 5.5 NORMAL APPROXIMATIONS TO BINOMIAL DISTRIBUTIONS

### 5.5 Try It Yourself Solutions

**1a.** $n = 100,\ p = 0.34,\ q = 0.66$

**b.** $np = 100(0.34) = 34,\ nq = 100(0.66) = 66$

**c.** Because $np \geq 5$ and $nq \geq 5$, the normal distribution can be used.

**d.** $\mu = np = 100(0.34) = 34$

$\sigma = \sqrt{npq} = \sqrt{100(0.34)(0.66)} \approx 4.74$

**2a.** (1) 57, 58, …, 83     (2) …, 52, 53, 54

**b.** (1) $56.5 < x < 83.5$     (2) $x < 54.5$

**3a.** $n = 100, \ p = 0.34, \ q = 0.66$

$np = 100(0.34) = 34, \ nq = 100(0.66) = 66$

Because $np \geq 5$ and $nq \geq 5$, the normal distribution can be used.

**b.** $\mu = np = 100(0.34) = 34$

$\sigma = \sqrt{npq} = \sqrt{100(0.34)(0.66)} \approx 4.74$

**c.** $P(x > 30.5)$

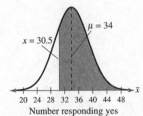

Number responding yes

**d.** $z = \dfrac{x - \mu}{\sigma} = \dfrac{30.5 - 34}{4.74} \approx -0.74$

**e.** $P(z < -0.74) = 0.2296$

$P(x > 30.5) = P(z > -0.74) = 1 - P(z < -0.74) = 1 - 0.2296 = 0.7704$

**4a.** $n = 200, \ p = 0.58$

$np = 200(0.58) = 116 \geq 5$ and $nq = 200(0.42) = 84 \geq 5$

The normal distribution can be used.

**b.** $\mu = np = 200(0.58) = 116$

$\sigma = \sqrt{npq} = \sqrt{100(0.58)(0.42)} \approx 6.98$

**c.** $P\left(x < 100.5\right)$

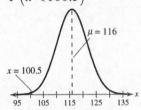

Number responding never

**d.** $z = \dfrac{x - \mu}{\sigma} \approx \dfrac{100.5 - 116}{6.98} \approx -2.22$

**e.** $P(x < 100.5) = P(z < -2.22) = 0.0132$

**5a.** $n = 75, \ p = 0.32$

$np = 75(0.32) = 24 \geq 5$ and $nq = 75(0.68) = 51 \geq 5$

The normal distribution can be used.

**b.** $\mu = np = 75(0.32) = 24$

$\sigma = \sqrt{npq} = \sqrt{75(0.32)(0.68)} \approx 4.04$

**c.** $P(14.5 < x < 15.5)$

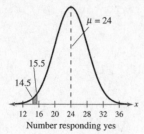

Number responding yes

**d.** $z = \dfrac{x - \mu}{\sigma} = \dfrac{14.5 - 24}{4.04} \approx -2.35$

$z = \dfrac{x - \mu}{\sigma} = \dfrac{15.5 - 24}{4.04} \approx -2.10$

**e.** $P(z < -2.35) = 0.0094$

$P(z < -2.10) = 0.0179$

$P(-2.35 < z < -2.10) = 0.0179 - 0.0094 = 0.0085$

## 5.5 EXERCISE SOLUTIONS

**1.** $np = (24)(0.85) = 20.4 \geq 5$

$nq = (24)(0.15) = 3.6 < 5$

Cannot use normal distribution.

**3.** $np = (18)(0.90) = 16.2 \geq 5$

$nq = (18)(0.10) = 1.8 < 5$

Cannot use normal distribution.

**5.** a        **7.** c

**9.** The probability of getting fewer than 25 successes; $P(x < 24.5)$

**11.** The probability of getting exactly 33 successes; $P(32.5 < x < 33.5)$

**13.** The probability of getting at most 150 successes; $P(x < 150.5)$

**15.** Binomial: $P(5 \leq x \leq 7) = P(x = 5) + P(x = 6) + P(x = 7)$

$$= {}_{16}C_5 (0.4)^5 (0.6)^{11} + {}_{16}C_6 (0.4)^6 (0.6)^{10} + {}_{16}C_7 (0.4)^7 (0.6)^9$$

$$\approx 0.549$$

Normal: $\mu = np = 16(0.4) = 6.4$, $\sigma = \sqrt{npq} = \sqrt{16(0.4)(0.6)} \approx 1.9596$

$z = \dfrac{x - \mu}{\sigma} \approx \dfrac{4.5 - 6.4}{1.9596} \approx -0.9696$

$z = \dfrac{x - \mu}{\sigma} \approx \dfrac{7.5 - 6.4}{1.9596} \approx 0.5613$

$P(5 \leq x \leq 7) \approx P(4.5 \leq x \leq 7.5) = P(-0.9696 \leq z \leq 0.5613)$

$$\approx 0.71272 - 0.16613 = 0.5466$$

(using technology)

The results are about the same.

**17.** $n = 30$, $p = 0.37$, $q = 0.63$

$np = 30(0.37) = 11.1 \geq 5$, $nq = 30(0.63) = 18.9 \geq 5$

Can use normal distribution.

$\mu = np = (30)(0.37) = 11.1$

$\sigma = \sqrt{npq} = \sqrt{(30)(0.37)(0.63)} \approx 2.644$

**19.** $n = 20$, $p = 0.78$, $q = 0.22$

$np = 20(0.78) = 15.6 \geq 5$, $nq = 20(0.22) = 4.4 < 5$

Cannot use normal distribution because $nq < 5$.

**21.** $n = 50$, $p = 0.65$, $q = 0.35$

$np = 50(0.65) = 32.5 \geq 5$, $nq = 50(0.35) = 17.5 \geq 5$

Can use normal distribution.

$\mu = np = (50)(0.65) = 32.5$

$\sigma = \sqrt{npq} = \sqrt{(50)(0.65)(0.35)} \approx 3.373$

**23.** $n = 100$, $p = 0.69$

$np = 100(0.69) = 69 \geq 5$, $nq = 100(0.31) = 31 \geq 5$

$\mu = np = 100(0.69) = 69$, $\sigma = \sqrt{npq} = \sqrt{100(0.69)(0.31)} \approx 4.6249$

Can use normal distribution.

(a) $z = \dfrac{x - \mu}{\sigma} \approx \dfrac{69.5 - 69}{4.625} \approx 0.1081$

$z = \dfrac{x - \mu}{\sigma} \approx \dfrac{70.5 - 69}{4.625} \approx 0.3243$

$P(x = 70) = P(69.5 < x < 70.5)$

$\approx P(0.1081 < z < 0.3243)$

$\approx 0.62716 - 0.54305 \approx 0.0841$ (using technology)

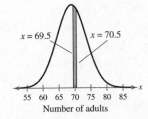

$x = 69.5$   $x = 70.5$

55 60 65 70 75 80 85
Number of adults

(b) $P(x \geq 70) = P(x > 69.5)$

$= P(z > 0.1081) = 1 - P(z < 0.1081)$ (using technology)

$\approx 1 - 0.5430 = 0.4570$

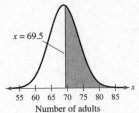

$x = 69.5$

55 60 65 70 75 80 85
Number of adults

(c) $P(x < 70) = P(x < 69.5) = P(z < 0.1081) \approx 0.5430$

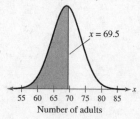

(d) No unusual events because all of the probabilities are greater than 0.05.

**25.** $n = 400$, $p = 0.08$

$np = 400(0.08) = 32 \geq 5$, $nq = 400(0.92) = 368 \geq 5$

$\mu = np = 400(0.08) = 32$, $\sigma = \sqrt{npq} = \sqrt{400(0.08)(0.92)} \approx 5.4259$

Can use normal distribution.

(a) $z = \dfrac{x - \mu}{\sigma} \approx \dfrac{40.5 - 32}{5.4259} \approx 1.5666$

$P(x \leq 40) \approx P(x < 40.5) = P(z < 1.5666) \approx 0.9414$ (using technology)

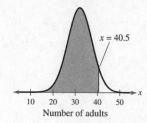

(b) $z = \dfrac{x - \mu}{\sigma} \approx \dfrac{50.5 - 32}{5.4259} \approx 3.4096$

$P(x > 50) \approx P(x > 50.5)$

$\approx P(z > 3.4096) = 1 - P(z < 3.4096)$ (using technology)

$\approx 1 - 0.9997 \approx 0.0003$

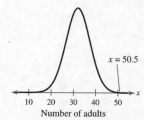

(c) $z = \dfrac{x - \mu}{\sigma} \approx \dfrac{19.5 - 32}{5.4259} \approx -2.3038$

$z = \dfrac{x - \mu}{\sigma} \approx \dfrac{30.5 - 32}{5.4259} \approx -0.2765$

$P(20 \leq x \leq 30) \approx P(19.5 < x < 30.5)$

$\approx P(-2.3038 < z < -0.2765)$ (using technology)

$\approx 0.39110 - 0.01062 \approx 0.3805$

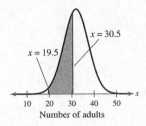

(d) The event in part (b) is unusual because its probability is less than 0.05.

**27.** $n = 14$, $p = 0.72$

$np = 14(0.72) = 10.08 \geq 5$, $nq = 14(0.28) = 3.92 < 5$

Cannot use normal distribution because $nq < 5$.

(a) $P(x = 8) = {}_{14}C_8 (0.72)^8 (0.28)^6 = 0.1045$

(b) $P(x \geq 10) = P(x = 10) + P(x = 11) + P(x = 12) + P(x = 13) + P(x = 14)$

$$= {}_{14}C_{10}(0.72)^{10}(0.28)^4 + {}_{14}C_{11}(0.72)^{11}(0.28)^3 + {}_{14}C_{12}(0.72)^{12}(0.28)^2$$

$$+ {}_{14}C_{13}(0.72)^{13}(0.28)^1 + {}_{14}C_{14}(0.72)^{14}(0.28)^0$$

$$\approx 0.2304 + 0.2154 + 0.1385 + 0.0548 + 0.0101 = 0.6491$$

(c) $P(x < 5) = P(x = 0) + P(x = 1) + P(x = 2) + P(x = 3) + P(x = 4)$

$$= {}_{14}C_0 (0.72)^0 (0.28)^{14} + {}_{14}C_1 (0.72)^1 (0.28)^{13} + {}_{14}C_2 (0.72)^2 (0.28)^{12}$$

$$+ {}_{14}C_3 (0.72)^3 (0.28)^{11} + {}_{14}C_4 (0.72)^4 (0.28)^{10}$$

$$\approx 0.0000 + 0.0000 + 0.0001 + 0.0001 + 0.0008 \approx 0.0009$$

(d) The event is part (c) is unusual because its probability is less than 0.05.

**29.** (a) $n = 250$, $p = 0.05$

$np = 250(0.05) = 12.5 \geq 5$, $nq = 250(0.95) = 237.5 \geq 5$

Can use normal distribution.

$\mu = np = 250(0.05) = 12.5$, $\sigma = \sqrt{npq} = \sqrt{250(0.05)(0.95)} \approx 3.446$

$P(x \geq 30) \approx P(x > 29.5)$

$$\approx P(z > 4.9333) = 1 - P(z < 4.9333) \text{ (using technology)}$$

$$\approx 1 - 1 \approx 0$$

(b) $n = 500$, $p = 0.05$

$np = 500(0.05) = 25 \geq 5$, $nq = 500(0.95) = 475 \geq 5$

Can use normal distribution.

$\mu = np = 500(0.05) = 25$, $\sigma = \sqrt{npq} = \sqrt{500(0.05)(0.95)} \approx 4.8734$

$P(x \geq 30) \approx P(x > 29.5)$

$$\approx P(z > 0.9234) = 1 - P(z < 0.9234) \text{ (using technology)}$$

$$\approx 1 - 0.8221 \approx 0.1779$$

(c) $n = 1000$, $p = 0.05$

$p = 1000(0.05) = 50 \geq 5$, $nq = 1000(0.95) = 950 \geq 5$

Can use normal distribution.

$\mu = np = 1000(0.05) = 50$, $\sigma = \sqrt{npq} = \sqrt{1000(0.05)(0.95)} \approx 6.892$

$$P(x \geq 30) \approx P(x > 29.5)$$
$$\approx P(z > -2.9745) = 1 - P(z < -2.9745) \text{ (using technology)}$$
$$\approx 1 - 0.0015 \approx 0.9985$$

**31.** $n = 250$, $p = 0.70$, $\mu = np = 250(0.70) = 175$, $\sigma = \sqrt{npq} = \sqrt{250(0.70)(0.30)} \approx 7.2457$

60% say no $\rightarrow n = 250(0.60) = 150$ say no while 100 say yes.

$$z = \frac{x - \mu}{\sigma} \approx \frac{100.5 - 175}{7.2457} \approx -10.28$$

$P(\text{less than or equal to 100 say yes}) = P(x \leq 100) \approx P(x < 100.5) \approx P(z < -10.28) \approx 0$

It is highly unlikely that 60% responded no. Answers will vary.

**33.** $n = 100$, $p = 0.75$, $\mu = np = 100(0.75) = 75$, $\sigma = \sqrt{npq} = \sqrt{100(0.75)(0.25)} \approx 4.3301$

$$z = \frac{x - \mu}{\sigma} \approx \frac{69.5 - 75}{4.3301} \approx -1.2702$$

$P(\text{reject claim}) = P(x < 70) \approx P(x < 69.5) \approx P(z < -1.2702) \approx 0.1020$

# CHAPTER 5 REVIEW EXERCISE SOLUTIONS

**1.** $\mu = 15$, $\sigma = 3$

**3.** Curve $B$ has the greatest mean because its line of symmetry occurs the farthest to the right.

**5.** 0.6772

**7.** 0.6293

**9.** $1 - 0.2843 = 0.7157$

**11.** 0.00236 (using technology)

**13.** $0.5 - 0.0505 = 0.4995$

**15.** $0.95637 - 0.51994 \approx 0.4364$ (using technology)

**17.** $0.0668 + 0.0668 = 0.1336$

**19.** $x = 17 \rightarrow z = \dfrac{x - \mu}{\sigma} = \dfrac{17 - 20.9}{5.2} = -.75$

$x = 29 \rightarrow z = \dfrac{x - \mu}{\sigma} = \dfrac{29 - 20.9}{5.2} \approx 1.56$

$x = 8 \rightarrow z = \dfrac{x - \mu}{\sigma} = \dfrac{8 - 20.9}{5.2} \approx -2.48$

$x = 23 \rightarrow z = \dfrac{x - \mu}{\sigma} = \dfrac{23 - 20.9}{5.2} \approx 0.40$

**21.** $P(z < 1.28) = 0.8997$

**23.** $P(-2.15 < x < 1.55) = 0.93943 - 0.01578 \approx 0.9237$ (using technology)

**25.** $P(z < -2.50 \text{ or } z > 2.50) = 2(0.0062) = 0.0124$

**27.** $z = \dfrac{x - \mu}{\sigma} = \dfrac{84 - 74}{8} = 1.25$

$P(x < 84) = P(z < 1.25) = 0.8944$

**29.** $z = \dfrac{x - \mu}{\sigma} = \dfrac{80 - 74}{8} = 0.75$

$P(x > 80) = P(z > 0.75) = 1 - P(z < 0.75) = 1 - 0.7734 = 0.2266$

**31.** $z = \dfrac{x - \mu}{\sigma} = \dfrac{60 - 74}{8} = -1.75$

$z = \dfrac{x - \mu}{\sigma} = \dfrac{70 - 74}{8} = -0.5$

$P(60 < x < 70) = P(-1.75 < z < -0.5) \approx 0.30854 - 0.04001 \approx 0.2685$ (using technology)

**33.** (a) $z = \dfrac{x - \mu}{\sigma} = \dfrac{200 - 267}{86} \approx -0.7791$

$P(x < 200) \approx P(z < -0.7791) \approx 0.2180$ (using technology)

(b) $z = \dfrac{x - \mu}{\sigma} = \dfrac{250 - 267}{86} \approx -0.1977$

$z = \dfrac{x - \mu}{\sigma} = \dfrac{350 - 267}{86} \approx 0.9651$

$P(250 < x < 350) \approx P(-0.1977 < z < 0.9651) \approx 0.83276 - 0.42165 \approx 0.4111$

(using technology)

(c) $z = \dfrac{x - \mu}{\sigma} = \dfrac{500 - 267}{86} \approx 2.7093$

$P(x > 500) \approx P(z > 2.7093) = 1 - P(z < 2.7093) \approx 1 - 0.9966 = 0.0034$ (using technology)

**35.** The event in part (c) is unusual because its probability is less than 0.05.

**37.** $z = -0.07$        **39.** $z = 2.457$ (using technology)        **41.** $z = 1.04$

**43.** 30.5% to right $\Rightarrow$ 69.5% to left, $z = 0.51$

**45.** $x = \mu + z\sigma = 127 + (-2.5)(3.81) = 117.48$ feet

**47.** 95th percentile $\Rightarrow$ Area $= 0.95 \Rightarrow z = 1.645$

$x = \mu + z\sigma = 127 + (1.645)(3.81) = 133.27$ feet

**49.** Top 10% $\Rightarrow$ Area $= 0.90 \Rightarrow z = 1.2816$ (using technology)

$x = \mu + z\sigma = 127 + (1.2816)(3.81) \approx 131.88$ feet

**51.**

| Sample | Mean | | Sample | Mean |
|--------|------|---|--------|------|
| 0, 0 | 0 | | 2, 0 | 1 |
| 0, 1 | 0.5 | | 2, 1 | 1.5 |
| 0, 2 | 1 | | 2, 2 | 2 |
| 0, 3 | 1.5 | | 2, 3 | 2.5 |
| 1, 0 | 0.5 | | 3, 0 | 1.5 |
| 1, 1 | 1 | | 3, 1 | 2 |
| 1, 2 | 1.5 | | 3, 2 | 2.5 |
| 1, 3 | 2 | | 3, 3 | 3 |

$\mu = 1.5$, $\sigma \approx 1.118$

$\mu_{\bar{x}} = 1.5$, $\sigma_{\bar{x}} \approx \dfrac{1.118}{\sqrt{2}} \approx 0.791$

The means are equal, but the standard deviation of the sampling distribution is smaller.

**53.** $\mu_{\bar{x}} = 85.6$, $\sigma_{\bar{x}} = \dfrac{\sigma}{\sqrt{n}} = \dfrac{20.5}{\sqrt{35}} \approx 3.465$

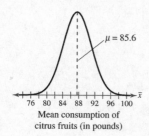

$\mu = 85.6$

76  80  84  88  92  96  100

Mean consumption of
citrus fruits (in pounds)

**55. (a)** $z = \dfrac{\bar{x} - \mu}{\dfrac{\sigma}{\sqrt{n}}} = \dfrac{200 - 267}{\dfrac{86}{\sqrt{12}}} \approx \dfrac{-67}{24.8261} \approx -2.6988$

$P(\bar{x} < 200) \approx P(z < -2.6988) \approx 0.0035$  (using technology)

**(b)** $z = \dfrac{\bar{x} - \mu}{\dfrac{\sigma}{\sqrt{n}}} = \dfrac{250 - 267}{\dfrac{86}{\sqrt{12}}} \approx \dfrac{-17}{24.8261} \approx -0.6848$

$z = \dfrac{\bar{x} - \mu}{\dfrac{\sigma}{\sqrt{n}}} = \dfrac{350 - 267}{\dfrac{86}{\sqrt{12}}} \approx \dfrac{83}{24.8261} \approx 3.3433$

$P(250 < \bar{x} < 350) \approx P(-0.6848 < z < 3.3433) \approx 0.99959 - 0.24675 \approx 0.7528$
(using technology)

**(c)** $z = \dfrac{\bar{x} - \mu}{\dfrac{\sigma}{\sqrt{n}}} = \dfrac{500 - 267}{\dfrac{86}{\sqrt{12}}} \approx \dfrac{233}{24.8261} \approx 9.39$

$P(\bar{x} > 500) = P(z > 9.39) = 1 - P(z < 9.39) \approx 1 - 1 \approx 0$  (using technology)

**(d)** The probabilities in parts (a) and (c) are smaller, and the probability in part (b) is larger. This is to be expected because the standard error of the sample mean is smaller.

**57. (a)** $z = \dfrac{\bar{x} - \mu}{\dfrac{\sigma}{\sqrt{n}}} = \dfrac{7200 - 6700}{\dfrac{1250}{\sqrt{36}}} \approx \dfrac{500}{208.3333} = 2.40$

$P(\bar{x} < 7200) = P(z < 2.40) = 0.9918$

**(b)** $z = \dfrac{\bar{x} - \mu}{\dfrac{\sigma}{\sqrt{n}}} = \dfrac{6500 - 6700}{\dfrac{1250}{\sqrt{36}}} \approx \dfrac{-200}{208.3333} = -0.96$

$P(\bar{x} > 6500) = P(z > -0.96) \approx 1 - 0.1685 = 0.8315$

**(c)** $z = \dfrac{\bar{x} - \mu}{\dfrac{\sigma}{\sqrt{n}}} = \dfrac{7000 - 6700}{\dfrac{1250}{\sqrt{36}}} \approx \dfrac{300}{208.3333} = 1.44$

$z = \dfrac{\bar{x} - \mu}{\dfrac{\sigma}{\sqrt{n}}} = \dfrac{7400 - 6700}{\dfrac{1250}{\sqrt{36}}} \approx \dfrac{700}{208.3333} = 3.36$

$P(7000 < \bar{x} < 7400) = P(1.44 < z < 3.36) \approx 0.99961 - 0.92507 \approx 0.0745$
(using technology)

**59. (a)** $z = \dfrac{\bar{x} - \mu}{\dfrac{\sigma}{\sqrt{n}}} = \dfrac{30,000 - 30,800}{\dfrac{5600}{\sqrt{45}}} \approx \dfrac{-800}{834.7987} \approx -0.9583$

$P(\bar{x} < 30,000) \approx P(z < -0.9583) \approx 0.1690$ (using technology)

**(b)** $z = \dfrac{\bar{x} - \mu}{\dfrac{\sigma}{\sqrt{n}}} = \dfrac{34,000 - 30,800}{\dfrac{5600}{\sqrt{45}}} \approx \dfrac{3200}{834.7987} \approx 3.8333$

$P(\bar{x} > 34,000) \approx P(z > 3.8333) \approx 1 - 0.9999 \approx 0.0001$ (using technology)

**61.** $n = 12$, $p = 0.73$, $q = 0.27$

$np = 12(0.73) = 8.76 \geq 5$, but $nq = 12(0.27) = 3.24 < 5$

Cannot use the normal distribution because $nq < 5$.

**63.** The probability of getting at least 25 successes; $P(x \geq 25) \approx P(x > 24.5)$

**65.** The probability of getting exactly 45 successes; $P(x = 45) \approx P(44.5 < x < 45.5)$

**67.** The probability of getting less than 60 successes; $P(x < 60) \approx P(x < 59.5)$

**69.** $n = 45$, $p = 0.52$, $q = 0.48$

$np = 45(.52) = 23.4 \geq 5$, $nq = 45(0.48) = 21.6 \geq 5$

Can use the normal distribution.

$\mu = np = 45(0.52) = 23.4$ $\sigma = \sqrt{npq} = \sqrt{45(0.52)(0.48)} \approx 3.3514$

(a) $z = \dfrac{x - \mu}{\sigma} \approx \dfrac{15.5 - 23.4}{3.3514} \approx -2.3572$

$P(x \leq 15) \approx P(x < 15.5) = P(z < -2.3572) \approx 0.0092$ (using technology)

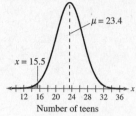

(b) $z = \dfrac{x - \mu}{\sigma} \approx \dfrac{25.5 - 23.4}{3.3514} \approx 0.6266$

$z = \dfrac{x - \mu}{\sigma} \approx \dfrac{24.5 - 23.4}{3.3514} \approx 0.3282$

$P(x = 25) \approx P(24.5 < x < 25.5) \approx P(0.3282 < z < 0.6266) \approx 0.73454 - 0.62863 \approx 0.1059$
(using technology)

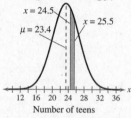

(c) $z = \dfrac{x - \mu}{\sigma} \approx \dfrac{30.5 - 23.4}{3.3514} \approx 2.1185$

$P(x > 30) \approx P(x > 30.5) \approx P(z > 2.1185) \approx 1 - 0.9829 \approx 0.0171$ (using technology)

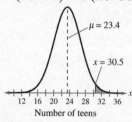

(d) The events in parts (a) and (c) are unusual because their probabilities are less than 0.05.

## CHAPTER 5 QUIZ SOLUTIONS

1. (a) $P(z > -2.54) = 1 - 0.0055 = 0.9945$

(b) $P(z < 3.09) = 0.9990$

(c) $P(-0.88 < z < 0.88) \approx 0.81057 - 0.18943 \approx 0.6211$ (using technology)

(d) $P(z < -1.445 \text{ or } z > -0.715) \approx 0.07423 + (1 - 0.23731) = 0.83692$ (using technology)

2. (a) $z = \dfrac{x - \mu}{\sigma} = \dfrac{5.97 - 9.2}{1.62} \approx -1.9938$

$P(x < 5.97) \approx P(z < -1.9938) \approx 0.0231$ (using technology)

(b) $z = \dfrac{x-\mu}{\sigma} = \dfrac{40.5-87}{19} \approx -2.4474$

$P(x > 40.5) \approx P(z > -2.4474) \approx 1 - 0.0072 \approx 0.9928$ (using technology)

(c) $z = \dfrac{x-\mu}{\sigma} = \dfrac{5.36-5.5}{0.08} = -1.75$

$z = \dfrac{x-\mu}{\sigma} = \dfrac{5.64-5.5}{0.08} = 1.75$

$P(5.36 < x < 5.64) = P(-1.75 < z < 1.75) \approx 0.95994 - 0.04006 \approx 0.9199$ (using technology)

(d) $z = \dfrac{x-\mu}{\sigma} = \dfrac{19.6-18.5}{4.25} \approx 0.2588$

$z = \dfrac{x-\mu}{\sigma} = \dfrac{26.1-18.5}{4.25} \approx 1.7882$

$P(19.6 < x < 26.1) \approx P(0.2588 < z < 1.7882) \approx 0.96313 - 0.60211 \approx 0.3610$
(using technology)

3. $z = \dfrac{x-\mu}{\sigma} = \dfrac{125-100}{15} \approx 1.6667$

$P(x > 125) \approx P(z > 1.6667) \approx 1 - 0.9522 \approx 0.0478$ (using technology)

Yes, the event is unusual because its probability is less than 0.05.

4. $z = \dfrac{x-\mu}{\sigma} = \dfrac{95-100}{15} \approx -0.3333$

$z = \dfrac{x-\mu}{\sigma} = \dfrac{105-100}{15} \approx 0.3333$

$P(95 < x < 105) \approx P(-0.3333 < z < 0.3333) \approx 0.63056 - 0.36944 \approx 0.2611$ (using technology)

No, the event is not unusual because its probability is greater than 0.05.

5. $z = \dfrac{x-\mu}{\sigma} = \dfrac{112-100}{15} = 0.80$

$P(x > 112) = P(z > 0.80) = 1 - .7881 = 0.2119 \rightarrow 21.19\%$ (using technology)

6. $z = \dfrac{x-\mu}{\sigma} = \dfrac{90-100}{15} \approx -0.6667$

$P(x < 90) = P(z < -0.6667) = 0.2525$ (using technology)

$(2000)(0.2525) = 505$ people

7. Top $5\% \rightarrow z \approx 1.6449$

$\mu + z\sigma = 100 + (1.6449)(15) \approx 124.7 \rightarrow 125$

8. Bottom $10\% \rightarrow z \approx -1.2816$

$\mu + z\sigma = 100 + (-1.2816)(15) \approx 80.8 \rightarrow 80$

(Because you are finding the highest score that would still place you in the bottom 10%, round down, because if you rounded up you would be outside the bottom 10%.)

9. $z = \dfrac{\bar{x} - \mu}{\dfrac{\sigma}{\sqrt{n}}} = \dfrac{105 - 100}{\dfrac{15}{\sqrt{60}}} \approx \dfrac{5}{1.9365} \approx 2.5820$

   $P(\bar{x} > 105) \approx P(z > 2.5820) \approx 1 - 0.9951 \approx 0.0049$ (using technology)

   About 0.5% of samples of 60 students will have a mean IQ score greater than 105. This is a very unusual event.

10. $z = \dfrac{x - \mu}{\sigma} = \dfrac{105 - 100}{15} \approx 0.3333$

    $P(x > 105) \approx P(z > 0.3333) \approx 1 - 0.6306 \approx 0.3694$ (using technology)

    $z = \dfrac{\bar{x} - \mu}{\dfrac{\sigma}{\sqrt{n}}} = \dfrac{105 - 100}{\dfrac{15}{\sqrt{15}}} \approx \dfrac{5}{3.873} \approx 1.2910$

    $P(\bar{x} > 105) \approx P(z > 1.2910) \approx 1 - 0.9016 \approx 0.0984$ (using technology)

    You are more likely to select one student with a test score greater than 105 because the standard error of the mean is less than the standard deviation.

11. $n = 45$, $p = 0.88$, $q = 0.12$

    $np = 45(0.88) = 39.6 \geq 5$, $nq = 45(0.12) = 5.4 \geq 5$

    Can use normal distribution.

    $\mu = np = 45(0.88) = 39.6$, $\sigma = \sqrt{npq} = \sqrt{45(0.88)(0.12)} \approx 2.180$

12. (a) $z = \dfrac{x - \mu}{\sigma} \approx \dfrac{35.5 - 39.6}{2.18} \approx -1.8807$

    $P(x \leq 35) \approx P(x < 35.5) \approx P(z < -1.8807) \approx 0.0300$ (using technology)

    (b) $z = \dfrac{x - \mu}{\sigma} \approx \dfrac{39.5 - 39.6}{2.18} \approx -0.0459$

    $P(x < 40) \approx P(x < 39.5) \approx P(z < -0.0459) \approx 0.4817$ (using technology)

    (c) $z = \dfrac{x - \mu}{\sigma} \approx \dfrac{42.5 - 39.6}{2.18} \approx 1.3303$

    $z = \dfrac{x - \mu}{\sigma} \approx \dfrac{43.5 - 39.6}{2.18} \approx 1.7890$

    $P(x = 43) \approx P(42.5 < x < 43.5) \approx P(1.3303 < z < 1.7890) \approx 0.96319 - 0.90829 \approx 0.0549$
    (using technology)

    (d) The event in part (a) is unusual because its probability is less than 0.05.

## CUMULATIVE REVIEW, CHAPTERS 3-5

1. (a) $np = 40(0.21) = 8.4 \geq 5$, $nq = 40(0.79) = 31.6 \geq 5$

    Can use normal distribution.

(b) $\mu = np = 40(0.21) = 8.4$, $\sigma = \sqrt{npq} = \sqrt{40(0.21)(0.79)} \approx 2.5760$

$z = \dfrac{x - \mu}{\sigma} \approx \dfrac{14.5 - 8.4}{2.576} \approx 2.3680$

$P(x \le 14) \approx P(x < 14.5) \approx P(z < 2.3680) \approx 0.9911 \,(\text{using technology})$

(c) $z = \dfrac{x - \mu}{\sigma} \approx \dfrac{13.5 - 8.4}{2.576} \approx 1.9798$

$P(x = 14) \approx P(13.5 < x < 14.5) \approx P(1.9798 < z < 2.3680) \approx 0.99106 - 0.97614 \approx 0.0149$

(using technology)

Yes, because the probability is less than 0.05.

**3.**

| $x$ | $P(x)$ | $xP(x)$ | $x - \mu$ | $(x - \mu)^2$ | $(x - \mu)^2 P(x)$ |
|---|---|---|---|---|---|
| 0 | 0.114 | 0.000 | -2.042 | 4.1698 | 0.4754 |
| 1 | 0.271 | 0.271 | -1.042 | 1.0858 | 0.2942 |
| 2 | 0.314 | 0.628 | -0.042 | 0.0018 | 0.0006 |
| 3 | 0.114 | 0.342 | 0.958 | 0.9178 | 0.1046 |
| 4 | 0.143 | 0.572 | 1.958 | 3.8338 | 0.5482 |
| 5 | 0.029 | 0.145 | 2.958 | 8.7498 | 0.2537 |
| 6 | 0.014 | 0.084 | 3.958 | 15.6658 | 0.2193 |
|  |  | $\sum xP(x) = 2.042$ |  |  | $\sum (x - \mu)^2 P(x) \approx 1.8961$ |

(a) $\mu = \sum xP(x) \approx 2.0$

(b) $\sigma^2 = \sum (x - \mu)^2 P(x) \approx 1.9$

(c) $\sigma = \sqrt{\sigma^2} \approx 1.4$

(d) $E(x) = \mu \approx 2.0$

(e) The number of fouls for a player in a game on average is bout 2 fouls. The standard deviation is 1.4, so games differ from the mean by no more than about 1 or 2 fouls.

**5.** (a) $(16)(15)(14)(13) = 43{,}680$

(b) $\dfrac{(7)(6)(5)(4)}{(16)(15)(14)(13)} = \dfrac{840}{43{,}680} \approx 0.0192$

**7.** 0.0010

**9.** $0.9984 - 0.500 = 0.4984$

**11.** $0.5478 + (1 - 0.9573) = 0.5905$

**13.** $p = \dfrac{1}{200} = 0.005$

(a) $P(x = 5) = (0.005)(0.995)^4 \approx 0.0049$

(b) $P(x \le 3) \approx (0.005)(0.995)^0 + (0.005)(0.995)^1 + (0.005)(0.995)^2$

$\approx 0.005 + 0.004975 + 0.004950 \approx 0.0149$

(c) $P(x > 20) = 1 - P(x \le 20) \approx 1 - 0.0954 = 0.9046$

**15. (a)** $\mu_{\bar{x}} = 70$

$$\sigma_{\bar{x}} = \frac{\sigma}{\sqrt{n}} = \frac{1.2}{\sqrt{40}} \approx 0.1897$$

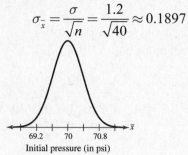

69.2    70    70.8
Initial pressure (in psi)

**(b)** $P(\bar{x} \leq 69) = P\left( z \leq \dfrac{69 - 70}{\dfrac{1.2}{\sqrt{15}}} \right) \approx P(z < -3.2275) \approx 0.0006$ (using technology)

**17. (a)** $_{12}C_4 = 495$

**(b)** $\dfrac{(1)(1)(1)(1)}{_{12}C_4} = 0.0020$

# Confidence Intervals

## 6.1 CONFIDENCE INTERVALS FOR THE MEAN (LARGE SAMPLES)

### 6.1 Try It Yourself Solutions

**1a.** $\bar{x} = \dfrac{\sum x}{n} = \dfrac{867}{30} = 28.9$

  **b.** A point estimate for the population mean number of hours worked is 28.9.

**2a.** $z_c = 1.96$, $n = 30$, $\sigma = 7.9$

  **b.** $E = z_c \dfrac{\sigma}{\sqrt{n}} \approx 1.96 \dfrac{7.9}{\sqrt{30}} \approx 2.8$

  **c.** You are 95% confident that the margin of error for the population mean is about 2.8 hours.

**3a.** $\bar{x} = 28.9$, $E \approx 2.8$

  **b.** $\bar{x} - E \approx 28.9 - 2.8 = 26.1$
    $\bar{x} + E \approx 28.9 + 2.8 = 31.7$

  **c.** With 95% confidence, you can say that the population mean number of hours worked is between 26.1 and 31.7 hours. This confidence interval is wider than the one found in Example 3.

**4a.** Enter the data.

  **b.** 75% CI: (28.2, 31.0)
    85% CI: (27.8, 31.4)
    90% CI: (27.5, 31.7)

  **c.** As the confidence level increases, so does the width of the interval.

**5a.** $n = 30$, $\bar{x} = 22.9$, $\sigma = 1.5$, $z_c = 1.645$

  **b.** $E = z_c \dfrac{\sigma}{\sqrt{n}} = 1.645 \dfrac{1.5}{\sqrt{30}} \approx 0.5$

    $\bar{x} - E \approx 22.9 - 0.5 = 22.4$
    $\bar{x} + E \approx 22.9 + 0.5 = 23.4$

  **c.** With 90% confidence, you can say that the mean age of the students is between 22.4 and 23.4 years. Because of the larger sample size, the confidence interval is slightly narrower.

**6a.** $z_c = 1.96$, $E = 2$, $\sigma \approx 7.9$

  **b.** $n = \left( \dfrac{z_c \sigma}{E} \right)^2 \approx \left( \dfrac{1.96 \cdot 7.9}{2} \right)^2 \approx 59.94 \to 60$

  **c.** You should have at least 60 employees in you sample. Because of the larger margin of error, the sample size needed is much smaller.

## 6.1 EXERCISE SOLUTIONS

1.  You are more likely to be correct using an interval estimate because it is unlikely that a point estimate will exactly equal the population mean.

3.  d; As the level of confidence increases, $z_c$ increases, causing wider intervals.

5.  1.28                    7.   1.15

9.  $\bar{x} - \mu = 3.8 - 4.27 = -0.47$            11.  $\bar{x} - \mu = 26.43 - 24.67 = 1.76$

13.  $E = z_c \dfrac{\sigma}{\sqrt{n}} = 1.96 \dfrac{5.2}{\sqrt{30}} \approx 1.861$            15.  $E = z_c \dfrac{\sigma}{\sqrt{n}} = 1.28 \dfrac{1.3}{\sqrt{75}} \approx 0.192$

17.  Because $c = 0.88$ is the lowest level of confidence, the interval associated with it will be the narrowest.  Thus, this matches (c).

19.  Because $c = 0.95$ is the third lowest level of confidence, the interval associated with it will be the third narrowest.  Thus, this matches (b).

21.  $\bar{x} \pm z_c \dfrac{\sigma}{\sqrt{n}} = 12.3 \pm 1.645 \dfrac{1.5}{\sqrt{50}} \approx 12.3 \pm 0.349 \approx (12.0,\ 12.6)$

23.  $\bar{x} \pm z_c \dfrac{\sigma}{\sqrt{n}} = 10.5 \pm 2.575 \dfrac{2.14}{\sqrt{45}} \approx 10.5 \pm 0.821 \approx (9.7,\ 11.3)$

25.  $(12.0, 14.8) \Rightarrow \bar{x} = \dfrac{14.8 + 12.0}{2} = 13.4,\ E = 14.8 - 13.4 = 1.4$

27.  $(1.71, 2.05) \Rightarrow \bar{x} = \dfrac{2.05 + 1.71}{2} = 1.88,\ E = 2.05 - 1.88 = 0.17$

29.  $c = 0.90 \Rightarrow z_c = 1.645$

$n = \left( \dfrac{z_c \sigma}{E} \right)^2 = \left( \dfrac{(1.645)(6.8)}{1} \right)^2 \approx 125.13 \Rightarrow 126$

31.  $c = 0.80 \Rightarrow z_c = 1.28$

$n = \left( \dfrac{z_c \sigma}{E} \right)^2 = \left( \dfrac{(1.28)(4.1)}{2} \right)^2 \approx 6.89 \Rightarrow 7$

33.  $(26.2, 30.1) \Rightarrow 2E = 30.1 - 26.2 = 3.9 \Rightarrow E = 1.95$ and $\bar{x} = 26.2 + E = 26.2 + 1.95 = 28.15$

**35.** 90% CI: $\bar{x} \pm z_c \dfrac{\sigma}{\sqrt{n}} = 3.63 \pm 1.645 \dfrac{0.21}{\sqrt{48}} \approx 3.63 \pm 0.0499 \approx (3.58,\ 3.68)$

95% CI: $\bar{x} \pm z_c \dfrac{\sigma}{\sqrt{n}} = 3.63 \pm 1.96 \dfrac{0.21}{\sqrt{48}} \approx 3.63 \pm 0.0594 \approx (3.57,\ 3.69)$

With 90% confidence, you can say that the population mean price is between \$3.58 and \$3.68. With 95% confidence, you can say that the population mean price is between \$3.57 and \$3.69. The 95% CI is wider.

**37.** $\bar{x} \pm z_c \dfrac{\sigma}{\sqrt{n}} = 2650 \pm 1.96 \dfrac{425}{\sqrt{50}} \approx 2650 \pm 117.80 \approx (2532.20,\ 2767.80)$

With 95% confidence, you can say that the population mean cost is between \$2532.20 and \$2767.80.

**39.** $\bar{x} \pm z_c \dfrac{\sigma}{\sqrt{n}} = 2650 \pm 1.96 \dfrac{425}{\sqrt{80}} \approx 2650 \pm 93.13 \approx (2556.9,\ 2743.1)$

The $n = 50$ CI is wider because a smaller sample is taken, giving less information about the population.

**41.** $\bar{x} \pm z_c \dfrac{\sigma}{\sqrt{n}} = 2650 \pm 1.96 \dfrac{375}{\sqrt{50}} \approx 2650 \pm 103.94 \approx (2546.06,\ 2753.94)$

The $\sigma = 425$ CI is wider because of the increased variability within the sample.

**43.** (a) An increase in the level of confidence will widen the confidence interval.
   (b) An increase in the sample size will narrow the confidence interval.
   (c) An increase in the standard deviation will widen the confidence interval.

**44.** Answers will vary.

**45.** $\bar{x} = \dfrac{\sum x}{n} = \dfrac{482}{20} = 24.1$

90% CI: $\bar{x} \pm z_c \dfrac{\sigma}{\sqrt{n}} = 24.1 \pm 1.645 \dfrac{4.3}{\sqrt{20}} \approx 24.1 \pm 1.582 \approx (22.5,\ 25.7)$

99% CI: $\bar{x} \pm z_c \dfrac{\sigma}{\sqrt{n}} = 24.1 \pm 2.575 \dfrac{4.3}{\sqrt{20}} \approx 24.1 \pm 2.476 \approx (21.6,\ 26.6)$

With 90% confidence, you can say that the population mean length of time is between 22.5 and 25.7 minutes. With 99% confidence, you can say that the population mean length of time is between 21.6 and 26.6 minutes. The 99% CI is wider.

**47.** $n = \left( \dfrac{z_c \sigma}{E} \right)^2 = \left( \dfrac{1.96 \cdot 4.8}{1} \right)^2 \approx 88.510 \to 89$

**49.** (a) $n = \left( \dfrac{z_c \sigma}{E} \right)^2 = \left( \dfrac{1.96 \cdot 2.8}{0.5} \right)^2 \approx 120.473 \to 121 \text{ servings}$

   (b) $n = \left( \dfrac{z_c \sigma}{E} \right)^2 = \left( \dfrac{2.575 \cdot 2.8}{0.5} \right)^2 \approx 207.936 \to 208 \text{ servings}$

(c) The 99% CI requires a larger sample because more information is needed from the population to be 99% confident.

**51.** (a) $n = \left( \dfrac{z_c \sigma}{E} \right)^2 = \left( \dfrac{1.645 \cdot 0.85}{0.25} \right)^2 \approx 31.282 \rightarrow 32 \text{ cans}$

(b) $n = \left( \dfrac{z_c \sigma}{E} \right)^2 = \left( \dfrac{1.645 \cdot 0.85}{0.15} \right)^2 \approx 86.893 \rightarrow 87 \text{ cans}$

(c) $E = 0.15$ requires a larger sample size. As the error size decreases, a larger sample must be taken to obtain enough information from the population to ensure the desired accuracy.

**53.** (a) $n = \left( \dfrac{z_c \sigma}{E} \right)^2 = \left( \dfrac{2.575 \cdot 0.25}{0.1} \right)^2 \approx 41.441 \rightarrow 42 \text{ soccer balls}$

(b) $n = \left( \dfrac{z_c \sigma}{E} \right)^2 = \left( \dfrac{2.575 \cdot 0.30}{0.1} \right)^2 \approx 59.676 \rightarrow 60 \text{ soccer balls}$

(c) $\sigma = 0.3$ requires a larger sample size. Due to the increased variability in the population, a larger sample is needed to ensure the desired accuracy.

**55.** (a) An increase in the level of confidence will increase the minimum sample size required.
(b) An increase (larger $E$) in the error tolerance will decrease the minimum sample size required.
(c) An increase in the population standard deviation will increase the minimum sample size required.

**57.** (a) $\sqrt{\dfrac{N-n}{N-1}} = \sqrt{\dfrac{1000-500}{1000-1}} \approx 0.707$     (b) $\sqrt{\dfrac{N-n}{N-1}} = \sqrt{\dfrac{1000-100}{1000-1}} \approx 0.949$

(c) $\sqrt{\dfrac{N-n}{N-1}} = \sqrt{\dfrac{1000-75}{1000-1}} \approx 0.962$     (d) $\sqrt{\dfrac{N-n}{N-1}} = \sqrt{\dfrac{1000-50}{1000-1}} \approx 0.975$

(e) The finite population correction factor approaches 1 as the sample size decreases and the population size remains the same.

**59.** (a) 99% CI: $\bar{x} \pm z_c \dfrac{\sigma}{\sqrt{n}} \sqrt{\dfrac{N-n}{N-1}} = 8.6 \pm 2.575 \dfrac{4.9}{\sqrt{25}} \sqrt{\dfrac{200-25}{200-1}} \approx 8.6 \pm 2.366 \approx (6.2,\ 11.0)$

(b) 90% CI: $\bar{x} \pm z_c \dfrac{\sigma}{\sqrt{n}} \sqrt{\dfrac{N-n}{N-1}} = 10.9 \pm 1.645 \dfrac{2.8}{\sqrt{50}} \sqrt{\dfrac{500-50}{500-1}} \approx 10.9 \pm 0.619 \approx (10.3,\ 11.5)$

(c) 95% CI: $\bar{x} \pm z_c \dfrac{\sigma}{\sqrt{n}} \sqrt{\dfrac{N-n}{N-1}} = 40.3 \pm 1.96 \dfrac{0.5}{\sqrt{68}} \sqrt{\dfrac{300-68}{300-1}} \approx 40.3 \pm 0.105 \approx (40.2,\ 40.4)$

(d) 80% CI: $\bar{x} \pm z_c \dfrac{\sigma}{\sqrt{n}} \sqrt{\dfrac{N-n}{N-1}} = 56.7 \pm 1.28 \dfrac{9.8}{\sqrt{36}} \sqrt{\dfrac{400-36}{400-1}} \approx 56.7 \pm 1.997 \approx (54.7,\ 58.7)$

## 6.2 CONFIDENCE INTERVALS FOR THE MEAN (SMALL SAMPLES)

### 6.2 Try It Yourself Solutions

**1a.** d.f. $= n - 1 = 22 - 1 = 21$

**b.** $c = 0.90$

**c.** $t_c = 1.721$

**d.** For a $t$-distribution curve with 21 degrees of freedom, 90% of the area under the curve lies between $t = \pm 1.721$.

**2a.** d.f. $= n - 1 = 16 - 1 = 15$

90% CI: $t_c = 1.753$

$$E = t_c \frac{s}{\sqrt{n}} = 1.753 \frac{10}{\sqrt{16}} \approx 4.4$$

99% CI: $t_c = 2.947$

$$E = t_c \frac{s}{\sqrt{n}} = 2.947 \frac{10}{\sqrt{16}} \approx 7.4$$

**b.** 90% CI: $\bar{x} \pm E \approx 162 \pm 4.4 = (157.6,\ 166.4)$

99% CI: $\bar{x} \pm E \approx 162 \pm 7.4 = (154.6,\ 169.4)$

**c.** With 90% confidence, you can say that the population mean temperature of coffee sold is between 157.6°F and 166.4°F.

With 99% confidence, you can say that the population mean temperature of coffee sold is between 154.6°F and 169.4°F.

**3a.** d.f. $= n - 1 = 36 - 1 = 35$

90% CI: $t_c = 1.690$

$$E = t_c \frac{s}{\sqrt{n}} = 1.690 \frac{2.39}{\sqrt{36}} \approx 0.67$$

95% CI: $t_c = 2.030$

$$E = t_c \frac{s}{\sqrt{n}} = 2.030 \frac{2.39}{\sqrt{36}} \approx 0.81$$

**b.** 90% CI: $\bar{x} \pm E \approx 9.75 \pm 0.67 = (9.08,\ 10.42)$

95% CI: $\bar{x} \pm E \approx 9.75 \pm 0.81 = (8.94,\ 10.56)$

**c.** With 90% confidence, you can say that the population mean number of days the car model sits on the lot is between 9.08 and 10.42 days.

With 95% confidence, you can say that the population mean number of days the car model sits on the lot is between 8.94 and 10.56 days. The 90% confidence interval is slightly narrower.

**4a.** Is $\sigma$ known? No

**b.** Is $n \geq 30$? No

Is the population normally distributed? Yes

**c.** Use the $t$-distribution because $\sigma$ is not known and the population is normally distributed.

## 6.2 EXERCISE SOLUTIONS

**1.** $t_c = 1.833$　　　**3.** $t_c = 2.947$

**5.** $E = t_c \dfrac{s}{\sqrt{n}} = 2.131 \dfrac{5}{\sqrt{16}} \approx 2.664$　　　**7.** $E = t_c \dfrac{s}{\sqrt{n}} = 1.691 \dfrac{2.4}{\sqrt{35}} \approx 0.686$

**9.** $\bar{x} \pm t_c \dfrac{s}{\sqrt{n}} = 12.5 \pm 2.015 \dfrac{2.0}{\sqrt{6}} \approx 12.5 \pm 1.645 \approx (10.9,\ 14.1)$

**11.** $\bar{x} \pm t_c \dfrac{s}{\sqrt{n}} = 4.3 \pm 2.650 \dfrac{0.34}{\sqrt{14}} \approx 4.3 \pm 0.241 \approx (4.1,\ 4.5)$

**13.** $(14.7,\ 22.1) \Rightarrow \bar{x} = \dfrac{14.7 + 22.1}{2} = 18.4 \Rightarrow E = 22.1 - 18.4 = 3.7$

**15.** $(64.6,\ 83.6) \Rightarrow \bar{x} = \dfrac{64.6 + 83.6}{2} = 74.1 \Rightarrow E = 83.6 - 74.1 = 9.5$

**17.** $E = t_c \dfrac{s}{\sqrt{n}} = 2.365 \dfrac{7.2}{\sqrt{8}} \approx 6.02$
$\bar{x} \pm E \approx 35.5 \pm 6.02 \approx (29.5,\ 41.5)$
With 95% confidence, you can say that the population mean commute time to work is between 29.5 and 41.5 minutes.

**19.** $E = t_c \dfrac{s}{\sqrt{n}} = 2.179 \dfrac{13.5}{\sqrt{13}} \approx 8.16$
$\bar{x} \pm E \approx 80 \pm 8.16 \approx (71.84,\ 88.16)$
With 95% confidence, you can say that the population mean repair cost is between $71.84 and $88.16.

**21.** $E = z_c \dfrac{\sigma}{\sqrt{n}} = 1.96 \dfrac{9.3}{\sqrt{8}} \approx 6.44$
$\bar{x} \pm E \approx 35.5 \pm 6.44 \approx (29.1,\ 41.9)$
With 95% confidence, you can say that the population mean commute time to work is between 29.1 and 41.9 minutes. This confidence interval is slightly wider than the one found in Exercise 17.

**23.** $E = z_c \dfrac{\sigma}{\sqrt{n}} = 1.96 \dfrac{15}{\sqrt{13}} \approx 8.15$
$\bar{x} \pm E \approx 80 \pm 8.15 \approx (71.85,\ 88.15)$
With 95% confidence, you can say that the population mean repair cost is between $71.85 and $88.15. This confidence interval is slightly narrower than the one found in Exercise 19.

**25.** (a) $\bar{x} \approx 1764.2$
(b) $s \approx 252.4$

(c) $\bar{x} \pm t_c \dfrac{s}{\sqrt{n}} \approx 1764.2 \pm 3.106 \dfrac{252.35}{\sqrt{12}} \approx 1764.2 \pm 226.26 \approx (1537.9, \ 1990.5)$

27. (a) $\bar{x} \approx 7.49$
    (b) $s \approx 1.64$
    (c) $\bar{x} \pm t_c \dfrac{s}{\sqrt{n}} \approx 7.49 \pm 2.947 \dfrac{1.64}{\sqrt{16}} \approx 7.49 \pm 1.21 \approx (6.28, \ 8.70)$

29. (a) $\bar{x} \approx 71,968.06$
    (b) $s \approx 15,426.35$
    (c) $\bar{x} \pm t_c \dfrac{s}{\sqrt{n}} \approx 71,968.06 \pm 2.441 \dfrac{15,426.35}{\sqrt{35}} \approx 71,968.06 \pm 6364.98 \approx (65,603.08, \ 78,333.04)$

31. Use a $t$-distribution because $\sigma$ unknown and $n \geq 30$ .

    $\bar{x} \pm t_c \dfrac{s}{\sqrt{n}} = 27.7 \pm 2.009 \dfrac{6.12}{\sqrt{50}} \approx 27.7 \pm 1.74 \approx (26.0, \ 29.4)$

    With 95% confidence, you can say that the population mean BMI is between 26.0 and 29.4.

33. Use a $t$-distribution because $\sigma$ unknown and $n \geq 30$ .

    $\bar{x} \pm t_c \dfrac{s}{\sqrt{n}} = 21.76 \pm 2.014 \dfrac{3.17}{\sqrt{45}} \approx 21.76 \pm 0.95 \approx (20.8, \ 22.7)$

    With 95% confidence, you can say that the population mean gas mileage is between 20.8 and 22.7.

35. Cannot use the standard normal distribution or the $t$-distribution because $\sigma$ unknown, $n < 30$, and we do not know if the times are normally distributed.

37. $n = 25, \bar{x} = 56.0, s = 0.25$

    $t = \dfrac{\bar{x} - \mu}{\dfrac{s}{\sqrt{n}}} = \dfrac{56.0 - 55.5}{\dfrac{0.25}{\sqrt{25}}} = 10$

    $-t_{0.99} = -2.797, \ t_{0.99} = 2.797$

    They are not making good tennis balls because for this sample the $t$-value is $t = 10$, which is not between $-t_{0.99} = -2.797$ and $t_{0.99} = 2.797$ .

## 6.3 CONFIDENCE INTERVALS FOR POPULATION PROPORTIONS

## 6.3 Try It Yourself Solutions

1a. $x = 123, n = 2462$

  b. $\hat{p} = \dfrac{123}{2462} \approx 5.0\%$

2a. $\hat{p} \approx 0.050, \hat{q} \approx 0.950$

**b.** $n\hat{p} \approx (2462)(0.050) = 123.1 > 5$

$n\hat{q} \approx (2462)(0.950) = 2338.9 > 5$

Distribution of $\hat{p}$ is approximately normal.

**c.** $z_c = 1.645$

$$E = z_c\sqrt{\frac{\hat{p}\hat{q}}{n}} \approx 1.645\sqrt{\frac{0.050 \cdot 0.950}{2462}} \approx 0.007$$

**d.** $\hat{p} \pm E \approx 0.050 \pm 0.007 \approx (0.043, \ 0.057)$

**e.** With 90% confidence, you can say that the proportion of U.S. teachers who say that "all or almost all" of the information they find using search engines online is accurate or trustworthy is between 4.3% and 5.7%.

**3a.** $n = 498, \ \hat{p} = 0.25$

$\hat{q} = 1 = \hat{p} - 1 = 0.25 \approx 0.75$

**b.** $n\hat{p} = 498 \cdot 0.25 = 124.5 > 5$

$n\hat{q} = 498 \cdot 0.75 = 373.5 > 5$

Distribution of $\hat{p}$ is approximately normal.

**c.** $z_c = 2.575, \ E = z_c\sqrt{\frac{\hat{p}\hat{q}}{n}} = 2.575\sqrt{\frac{0.25(0.75)}{498}} \approx 0.050$

**d.** $\hat{p} \pm E \approx 0.25 \pm 0.050 = (0.20, \ 0.30)$

**e.** With 99% confidence, you can say that the proportion of U.S. adults who think that people over 65 are the more dangerous drivers is between 20% and 30%.

**4a.** (1) $\hat{p} = 0.5, \ \hat{q} = 0.5, \ z_c = 1.645, \ E = 0.02$

(2) $\hat{p} = 0.31, \ \hat{q} = 0.69, \ z_c = 1.645, \ E = 0.02$

**b.** (1) $n = \hat{p}\hat{q}\left(\frac{z_c}{E}\right)^2 = (0.5)(0.5)\left(\frac{1.645}{0.02}\right)^2 \approx 1691.266 \rightarrow 1692$

(2) $n = \hat{p}\hat{q}\left(\frac{z_c}{E}\right)^2 = (0.31)(0.69)\left(\frac{1.645}{0.02}\right)^2 \approx 1447.05 \rightarrow 1448$

**c.** (1) At least 1692 adults should be included in the sample.

(2) At least 1448 adults should be included in the sample.

## 6.3 EXERCISE SOLUTIONS

**1.** False. To estimate the value of $p$, the population proportion of successes, use the point estimate

$\hat{p} = \frac{x}{n}$.

**3.** $\hat{p} = \frac{x}{n} = \frac{662}{1002} \approx 0.661, \ \hat{q} = 1 - \hat{p} \approx 0.339$      **5.** $\hat{p} = \frac{x}{n} = \frac{4912}{11,605} \approx 0.423, \ \hat{q} = 1 - \hat{p} \approx 0.577$

7. $(0.905, 0.933) \rightarrow \hat{p} = \dfrac{0.905 + 0.933}{2} = 0.919 \Rightarrow E = 0.933 - 0.919 = 0.014$

9. $(0.512, 0.596) \rightarrow \hat{p} = \dfrac{0.512 + 0.596}{2} = 0.554 \Rightarrow E = 0.596 - 0.554 = 0.042$

11. $\hat{p} = \dfrac{x}{n} = \dfrac{396}{674} \approx 0.588,\ \hat{q} = 1 - \hat{p} \approx 0.412$

   90% CI: $\hat{p} \pm z_c \sqrt{\dfrac{\hat{p}\hat{q}}{n}} \approx 0.588 \pm 1.645 \sqrt{\dfrac{(0.588)(0.412)}{674}} \approx 0.588 \pm 0.031 \approx (0.557,\ 0.619)$

   95% CI: $\hat{p} \pm z_c \sqrt{\dfrac{\hat{p}\hat{q}}{n}} \approx 0.588 \pm 1.96 \sqrt{\dfrac{(0.588)(0.412)}{674}} \approx 0.588 \pm 0.037 \approx (0.551,\ 0.625)$

   With 90% confidence, you can say that the population proportion of U.S. males ages 18-64 who say they have gone to the dentist in the past year is between 55.7% and 61.9%. With 95% confidence, you can say it is between 55.1% and 62.5%. The 95% confidence interval is slightly wider.

13. $\hat{p} = \dfrac{x}{n} = \dfrac{1435}{3110} \approx 0.461,\ \hat{q} = 1 - \hat{p} \approx 0.539$

   $\hat{p} \pm z_c \sqrt{\dfrac{\hat{p}\hat{q}}{n}} \approx 0.461 \pm 2.575 \sqrt{\dfrac{(0.461)(0.539)}{3110}} \approx 0.461 \pm 0.023 \approx (0.438,\ 0.484)$

   With 99% confidence, you can say that the population proportion of U.S. adults who say they have started paying bills online in the past year is between 43.8% and 48.4%.

15. $\hat{p} = \dfrac{x}{n} = \dfrac{1272}{2230} \approx 0.570,\ \hat{q} = 1 - \hat{p} = 0.430$

   $\hat{p} \pm z_c \sqrt{\dfrac{\hat{p}\hat{q}}{n}} \approx 0.570 \pm 1.96 \sqrt{\dfrac{(0.570)(0.430)}{2230}} \approx 0.570 \pm 0.021 \approx (0.549,\ 0.591)$

17. (a) $n = \hat{p}\hat{q}\left(\dfrac{z_c}{E}\right)^2 = 0.5 \cdot 0.5 \left(\dfrac{1.96}{0.04}\right)^2 \approx 600.25 \rightarrow 601$ adults

   (b) $n = \hat{p}\hat{q}\left(\dfrac{z_c}{E}\right)^2 = 0.48 \cdot 0.52 \left(\dfrac{1.96}{0.04}\right)^2 \approx 599.3 \rightarrow 600$ adults

   (c) Having an estimate of the population proportion reduces the minimum sample size needed.

19. (a) $n = \hat{p}\hat{q}\left(\dfrac{z_c}{E}\right)^2 = 0.5 \cdot 0.5 \left(\dfrac{1.645}{0.03}\right)^2 \approx 751.67 \rightarrow 752$ adults

   (b) $n = \hat{p}\hat{q}\left(\dfrac{z_c}{E}\right)^2 = 0.43 \cdot 0.57 \left(\dfrac{1.645}{0.03}\right)^2 \approx 736.94 \rightarrow 737$ adults

   (c) Having an estimate of the population proportion reduces the minimum sample size needed.

**21.** (a) $\hat{p} = 0.69$, $\hat{q} = 0.31$, $n = 1044$

$$\hat{p} \pm z_c \sqrt{\frac{\hat{p}\hat{q}}{n}} = 0.69 \pm 2.575 \sqrt{\frac{(0.69)(0.31)}{1044}} \approx 0.69 \pm 0.037 \approx (0.653, \ 0.727)$$

(b) $\hat{p} = 0.72$, $\hat{q} = 0.28$, $n = 871$

$$\hat{p} \pm z_c \sqrt{\frac{\hat{p}\hat{q}}{n}} = 0.72 \pm 2.575 \sqrt{\frac{(0.72)(0.28)}{871}} \approx 0.72 \pm 0.039 \approx (0.681, \ 0.759)$$

(c) $\hat{p} = 0.62$, $\hat{q} = 0.38$, $n = 1097$

$$\hat{p} \pm z_c \sqrt{\frac{\hat{p}\hat{q}}{n}} = 0.62 \pm 2.575 \sqrt{\frac{(0.62)(0.38)}{1097}} \approx 0.62 \pm 0.038 \approx (0.582, \ 0.658)$$

(d) $\hat{p} = 0.75$, $\hat{q} = 0.25$, $n = 1003$

$$\hat{p} \pm z_c \sqrt{\frac{\hat{p}\hat{q}}{n}} = 0.75 \pm 2.575 \sqrt{\frac{(0.75)(0.25)}{1003}} \approx 0.75 \pm 0.035 \approx (0.715, \ 0.785)$$

**23.** (a) $\hat{p} = 0.32$, $\hat{q} = 1 - \hat{p} = 0.68$

$$\hat{p} \pm z_c \sqrt{\frac{\hat{p}\hat{q}}{n}} = 0.32 \pm 1.96 \sqrt{\frac{(0.32)(0.68)}{400}} \approx 0.32 \pm 0.046 \approx (0.274, \ 0.366)$$

(b) $\hat{p} = 0.56$, $\hat{q} = 1 - \hat{p} = 0.44$

$$\hat{p} \pm z_c \sqrt{\frac{\hat{p}\hat{q}}{n}} = 0.56 \pm 1.96 \sqrt{\frac{(0.56)(0.44)}{400}} \approx 0.56 \pm 0.049 \approx (0.511, \ 0.609)$$

**25.** No, it is unlikely that the two proportions are equal because the confidence intervals estimating the proportions do not overlap. The 99% confidence intervals are (0.260, 0.380) and (0.496, 0.624). Although these intervals are wider, they still do not overlap.

**27.** $31.4\% \pm 1\% \rightarrow (30.4\%, \ 32.4\%) \rightarrow (0.304, \ 0.324)$

$$E = z_c \sqrt{\frac{\hat{p}\hat{q}}{n}} \rightarrow z_c = E\sqrt{\frac{n}{\hat{p}\hat{q}}} = 0.01\sqrt{\frac{8451}{(0.314)(0.686)}} \approx 1.981 \rightarrow z_c = 1.98$$

$P(-1.98 < z < 1.98) = 0.9762 - 0.0238 = 0.9524 = c$

(30.4%, 32.4%) is approximately a 95.2% CI.

**29.** If $n\hat{p} < 5$ or $n\hat{q} < 5$, the sampling distribution of $\hat{p}$ may not be normally distributed, so $z_c$ cannot be used to calculate the confidence interval.

**31.**

| $\hat{p}$ | $\hat{q}=1-\hat{p}$ | $\hat{p}\hat{q}$ | | $\hat{p}$ | $\hat{q}=1-\hat{p}$ | $\hat{p}\hat{q}$ |
|---|---|---|---|---|---|---|
| 0.0 | 1.0 | 0.00 | | 0.45 | 0.55 | 0.2475 |
| 0.1 | 0.9 | 0.09 | | 0.46 | 0.54 | 0.2484 |
| 0.2 | 0.8 | 0.16 | | 0.47 | 0.53 | 0.2491 |
| 0.3 | 0.7 | 0.21 | | 0.48 | 0.52 | 0.2496 |
| 0.4 | 0.6 | 0.24 | | 0.49 | 0.51 | 0.2499 |
| 0.5 | 0.5 | 0.25 | | 0.50 | 0.50 | 0.2500 |
| 0.6 | 0.4 | 0.24 | | 0.51 | 0.49 | 0.2499 |
| 0.7 | 0.3 | 0.21 | | 0.52 | 0.48 | 0.2496 |
| 0.8 | 0.2 | 0.16 | | 0.53 | 0.47 | 0.2491 |
| 0.9 | 0.1 | 0.09 | | 0.54 | 0.46 | 0.2484 |
| 1.0 | 0.0 | 0.00 | | 0.55 | 0.45 | 0.2475 |

$\hat{p}=0.5$ gives the maximum value of $\hat{p}\hat{q}$.

## 6.4 CONFIDENCE INTERVALS FOR VARIANCE AND STANDARD DEVIATION

## 6.4 Try It Yourself Solutions

**1a.** d.f. $= n-1 = 29$
level of confidence $= 0.90$

**b.** Area to the right of $\chi_R^2$ is 0.05.

Area to the right of $\chi_L^2$ is 0.95.

**c.** $\chi_R^2 = 42.557$, $\chi_L^2 = 17.708$

**d.** For a chi-square distribution curve with 29 degrees of freedom, 90% of the area under the curve lies between 17.708 and 42.557.

**2a.** 90% CI: $\chi_R^2 = 42.557$, $\chi_L^2 = 17.708$

95% CI: $\chi_R^2 = 45.722$, $\chi_L^2 = 16.047$

**b.** 90% CI for $\sigma^2$ : $\left( \dfrac{(n-1)s^2}{\chi_R^2}, \dfrac{(n-1)s^2}{\chi_L^2} \right) = \left( \dfrac{29 \cdot (1.2)^2}{42.557}, \dfrac{29 \cdot (1.2)^2}{17.708} \right) \approx (0.98,\ 2.36)$

95% CI for $\sigma^2$ : $\left( \dfrac{(n-1)s^2}{\chi_R^2}, \dfrac{(n-1)s^2}{\chi_L^2} \right) = \left( \dfrac{29 \cdot (1.2)^2}{45.722}, \dfrac{29 \cdot (1.2)^2}{16.047} \right) \approx (0.91,\ 2.60)$

**c.** 90% CI for $\sigma$ : $\left( \sqrt{0.981},\ \sqrt{2.358} \right) \approx (0.99,\ 1.54)$

95% CI for $\sigma$ : $\left( \sqrt{0.913},\ \sqrt{2.602} \right) \approx (0.96,\ 1.61)$

**d.** With 90% confidence, you can say that the population variance is between 0.98 and 2.36 and that the population standard deviation is between 0.99 and 1.54. With 95% confidence, you can say that the population variance is between 0.91 and 2.60, and that the population standard deviation is between 0.96 and 1.61.

## 6.4 EXERCISE SOLUTIONS

**1.** Yes.

**3.** $\chi_R^2 = 14.067$, $\chi_L^2 = 2.167$  **5.** $\chi_R^2 = 32.852$, $\chi_L^2 = 8.907$  **7.** $\chi_R^2 = 52.336$, $\chi_L^2 = 13.121$

**9.** (a) $\left( \dfrac{(n-1)s^2}{\chi_R^2}, \dfrac{(n-1)s^2}{\chi_L^2} \right) \approx \left( \dfrac{29 \cdot (11.56)}{45.722}, \dfrac{29 \cdot (11.56)}{16.047} \right) \approx (7.33,\ 20.89)$

(b) $\left( \sqrt{7.3321},\ \sqrt{20.8911} \right) \approx (2.71,\ 4.57)$

**11.** (a) $\left( \dfrac{(n-1)s^2}{\chi_R^2}, \dfrac{(n-1)s^2}{\chi_L^2} \right) \approx \left( \dfrac{17 \cdot (35)^2}{27.587}, \dfrac{17 \cdot (35)^2}{8.672} \right) \approx (755,\ 2401)$

(b) $\left( \sqrt{754.885},\ \sqrt{2401.407} \right) \approx (27,\ 49)$

**13.** (a) $s^2 \approx 0.0793$

$\left( \dfrac{(n-1)s^2}{\chi_R^2}, \dfrac{(n-1)s^2}{\chi_L^2} \right) \approx \left( \dfrac{16 \cdot (0.0793)}{28.845}, \dfrac{16 \cdot (0.0793)}{6.908} \right) \approx (0.0440,\ 0.1837)$

(b) $\left( \sqrt{0.04399},\ \sqrt{0.18367} \right) \approx (0.2097,\ 0.4286)$

With 95% confidence, you can say that the population variance is between 0.0440 and 0.1837, and the population standard deviation is between 0.2097 and 0.4286 inch.

**15.** (a) $s^2 \approx 0.06414$

$\left( \dfrac{(n-1)s^2}{\chi_R^2}, \dfrac{(n-1)s^2}{\chi_L^2} \right) \approx \left( \dfrac{17 \cdot (0.06414)}{35.718}, \dfrac{17 \cdot (0.06414)}{5.697} \right) \approx (0.0305,\ 0.1914)$

(b) $\left( \sqrt{0.030527},\ \sqrt{0.191395} \right) \approx (0.1747,\ 0.4375)$

With 99% confidence, you can say that the population variance is between 0.0305 and 0.1914, and the population standard deviation is between 0.1747 and 0.4375 hour.

**17.** (a) $\left( \dfrac{(n-1)s^2}{\chi_R^2}, \dfrac{(n-1)s^2}{\chi_L^2} \right) \approx \left( \dfrac{13 \cdot (3.90)^2}{29.819}, \dfrac{13 \cdot (390)^2}{3.565} \right) \approx (6.63,\ 55.46)$

(b) $\left( \sqrt{6.631},\ \sqrt{55.464} \right) \approx (2.58,\ 7.45)$

With 99% confidence, you can say that the population variance is between 6.63 and 55.46, and the population standard deviation is between \$2.58 and \$7.55.

**19.** (a) $\left( \dfrac{(n-1)s^2}{\chi_R^2}, \dfrac{(n-1)s^2}{\chi_L^2} \right) \approx \left( \dfrac{18 \cdot (15)^2}{31.526}, \dfrac{18 \cdot (15)^2}{8.231} \right) \approx (128,\ 492)$

(b) $\left( \sqrt{128.465},\ \sqrt{492.042} \right) \approx (11,\ 22)$

With 95% confidence, you can say that the population variance is between 128 and 492, and the population standard deviation is between 11 and 22 grains per gallon.

**21.** (a) $\left(\dfrac{(n-1)s^2}{\chi_R^2}, \dfrac{(n-1)s^2}{\chi_L^2}\right) \approx \left(\dfrac{13 \cdot (3725)^2}{19.812}, \dfrac{13 \cdot (3725)^2}{7.042}\right) \approx (9{,}104{,}741, \ 25{,}615{,}326)$

(b) $\left(\sqrt{9{,}104{,}741}, \ \sqrt{25{,}615{,}326}\right) \approx (3017, \ 5061)$

With 80% confidence, you can say that the population variance is between 9,104,741 and 25,615,326, and the population standard deviation is between \$3017 and \$5061.

**23.** (a) $\left(\dfrac{(n-1)s^2}{\chi_R^2} \cdot \dfrac{(n-1)s^2}{\chi_L^2}\right) \approx \left(\dfrac{(21)(3.6)^2}{38.932} \cdot \dfrac{(21)(3.6)^2}{8.897}\right) \approx (7.0, \ 30.6)$

(b) $\left(\sqrt{6.99}, \ \sqrt{30.59}\right) \approx (2.6, \ 5.5)$

With 98% confidence, you can say that the population variance is between 7.0 and 30.6, and the population standard deviation is between 2.6 and 5.5 minutes.

**25.** 95% CI for $\sigma$: $(0.2097, \ 0.4286)$

Yes, because all of the values in the confidence interval are less than 0.5.

**27.** Answers will vary. *Sample answer:* Unlike a confidence interval for a population mean or proportion, a confidence interval for a population variance does not have a margin of error. The left and right endpoints must be calculated separately.

## CHAPTER 6 REVIEW EXERCISE SOLUTIONS

**1.** (a) $\bar{x} \approx 103.5$

(b) $E = z_c \dfrac{\sigma}{\sqrt{n}} \approx 1.645 \dfrac{45}{\sqrt{40}} \approx 11.7$

**3.** $\bar{x} \pm z_c \dfrac{\sigma}{\sqrt{n}} = 103.5 \pm 1.645 \dfrac{45}{\sqrt{40}} \approx 103.5 \pm 11.7 \approx (91.8, \ 115.2)$

With 90% confidence, you can say that the population mean waking time is between 91.8 and 115.2 minutes past 5:00 A.M.

**5.** $(20.75, \ 24.10) \rightarrow \bar{x} = \dfrac{20.75 + 24.10}{2} = 22.425 \rightarrow E = 24.10 - 22.425 = 1.675$

**7.** $n = \left(\dfrac{z_c \sigma}{E}\right)^2 \approx \left(\dfrac{1.96 \cdot 45}{10}\right)^2 \approx 77.79 \Rightarrow 78$ people

**9.** $t_c = 1.383$      **11.** $t_c = 2.624$

**13.** $E = t_c \dfrac{s}{\sqrt{n}} = 1.753 \dfrac{25.6}{\sqrt{16}} \approx 11.2$      **15.** $E = t_c \dfrac{s}{\sqrt{n}} = 2.718 \left(\dfrac{0.9}{\sqrt{12}}\right) \approx 0.7$

**17.** $\bar{x} \pm t_c \dfrac{s}{\sqrt{n}} = 72.1 \pm 1.753 \dfrac{25.6}{\sqrt{16}} \approx 72.1 \pm 11.2 \approx (60.9,\ 83.3)$

**19.** $\bar{x} \pm t_c \dfrac{s}{\sqrt{n}} = 6.8 \pm 2.718 \left( \dfrac{0.9}{\sqrt{12}} \right) \approx 6.8 \pm 0.7 \approx (6.1,\ 7.5)$

**21.** $\bar{x} \pm t_c \dfrac{s}{\sqrt{n}} = 2929 \pm 1.703 \dfrac{786}{\sqrt{28}} \approx 2929 \pm 252.96 \approx (2676.0,\ 3182.0)$

With 90% confidence, you can say that the population mean annual fuel cost is between \$2676 and \$3182.

**23.** $\hat{p} = \dfrac{x}{n} = \dfrac{375}{814} \approx 0.461,\ \hat{q} = 1 - \hat{p} \approx 0.539$   **25.** $\hat{p} = \dfrac{x}{n} = \dfrac{552}{1023} \approx 0.540,\ \hat{q} = 1 - \hat{p} \approx 0.460$

**27.** $\hat{p} \pm z_c \sqrt{\dfrac{\hat{p}\hat{q}}{n}} = 0.461 \pm 1.96 \sqrt{\dfrac{0.461 \cdot 0.539}{814}} \approx 0.461 \pm 0.034 \approx (0.427,\ 0.495)$

With 95% confidence, you can say that the population proportion of U.S. adults who say the economy is the most important issue facing the country today is between 42.7% and 49.5%.

**29.** $\hat{p} \pm z_c \sqrt{\dfrac{\hat{p}\hat{q}}{n}} \approx 0.540 \pm 1.645 \sqrt{\dfrac{0.540 \cdot 0.460}{1023}} \approx 0.540 \pm 0.026 \approx (0.514,\ 0.566)$

With 90% confidence, you can say that the population proportion of U.S. adults who say they have worked the night shift at some point in their lives is between 51.4% and 56.6%.

**31.** (a) $n = \hat{p}\hat{q} \left( \dfrac{z_c}{E} \right)^2 = 0.50 \cdot 0.50 \left( \dfrac{1.96}{0.05} \right)^2 \approx 384.16 \rightarrow 385$ adults

  (b) $n = \hat{p}\hat{q} \left( \dfrac{z_c}{E} \right)^2 = 0.63 \cdot 0.37 \left( \dfrac{1.96}{0.05} \right)^2 \approx 358.19 \rightarrow 359$ adults

  (c) The minimum sample size needed is smaller when a preliminary estimate is available.

**33.** $\chi_R^2 = 23.337$, $\chi_L^2 = 4.404$   **35.** $\chi_R^2 = 24.996$, $\chi_L^2 = 7.261$

**37.** $s^2 \approx 49.0294$

  (a) 95% CI for $\sigma^2$: $\left( \dfrac{(n-1)s^2}{\chi_R^2},\ \dfrac{(n-1)s^2}{\chi_L^2} \right) \approx \left( \dfrac{16 \cdot (49.0294)}{28.845},\ \dfrac{16 \cdot (49.0294)}{6.908} \right) \approx (27.2,\ 113.6)$

  (b) 95% CI for $\sigma$: $\left( \sqrt{27.195},\ \sqrt{113.560} \right) \approx (5.2,\ 10.7)$

   With 95% confidence we can say that the population variance is between 27.2 and 113.6, and the population standard deviation is between 5.2 and 10.7 ounces.

## CHAPTER 6 QUIZ SOLUTIONS

1.  (a) $\bar{x} \approx 6.848$

    (b) $E = z_c \dfrac{\sigma}{\sqrt{n}} \approx 1.96 \dfrac{2.4}{\sqrt{30}} \approx 0.859$

    (c) $\bar{x} \pm z_c \dfrac{\sigma}{\sqrt{n}} \approx 6.848 \pm 1.96 \dfrac{2.4}{\sqrt{30}} \approx 6.848 \pm 0.859 \approx (5.989,\ 7.707)$

    With 95% confidence, you can say that the population mean amount of time is between 5.989 and 7.707 minutes.

2.  $n = \left(\dfrac{z_c \sigma}{E}\right)^2 = \left(\dfrac{2.575 \cdot 2.4}{1}\right)^2 \approx 38.18 \rightarrow 39$ students

3.  (a) $\bar{x} = 6.61,\ s \approx 3.376$

    (b) $\bar{x} \pm t_c \dfrac{s}{\sqrt{n}} \approx 6.61 \pm 1.833 \dfrac{3.376}{\sqrt{10}} \approx 6.61 \pm 1.957 \approx (4.65,\ 8.57)$

    With 90% confidence you can say that the population mean amount of time is between 4.65 and 8.57 minutes.

    (c) $\bar{x} \pm z_c \dfrac{\sigma}{\sqrt{n}} \approx 6.61 \pm 1.645 \dfrac{3.5}{\sqrt{10}} \approx 6.61 \pm 1.82 \approx (4.79,\ 8.43)$

    With 90% confidence you can say that the population mean amount of time is between 4.79 and 8.43 minutes. This confidence interval is narrower than the one found in part (b).

4.  $\bar{x} \pm t_c \dfrac{s}{\sqrt{n}} = 31{,}721 \pm 2.201 \dfrac{5260}{\sqrt{12}} \approx 31{,}721 \pm 3342 \approx (28{,}379,\ 35{,}063)$

    With 95% confidence you can say that the population mean annual earnings is between $28,379 and $35,063.

5.  (a) $\hat{p} = \dfrac{x}{n} = \dfrac{779}{1022} \approx 0.762$

    (b) $\hat{p} \pm z_c \sqrt{\dfrac{\hat{p}\hat{q}}{n}} \approx 0.762 \pm 1.645 \sqrt{\dfrac{0.762 \cdot 0.238}{1022}} \approx 0.762 \pm 0.022 \approx (0.740,\ 0.784)$

    With 90% confidence you can say that the population proportion of U.S. adults who think that the United States should not put more emphasis on producing domestic energy from solar power is between 74.0% and 78.4%.

    (c) $n = \hat{p}\hat{q}\left(\dfrac{z_c}{E}\right)^2 \approx 0.762 \cdot 0.238 \left(\dfrac{2.575}{0.04}\right)^2 \approx 751.56 \rightarrow 752$ adults

6.  (a) $\left(\dfrac{(n-1)s^2}{\chi_R^2},\ \dfrac{(n-1)s^2}{\chi_L^2}\right) = \left(\dfrac{9 \cdot (3.38)^2}{19.023},\ \dfrac{9 \cdot (3.38)^2}{2.700}\right) \approx (5.41,\ 38.08)$

    (b) $\left(\sqrt{5.4050},\ \sqrt{38.0813}\right) \approx (2.32,\ 6.17)$

    With 95% confidence you can say that the population standard deviation is between 2.32 and 6.17 minutes.

# Hypothesis Testing with One Sample

## 7.1 INTRODUCTION TO HYPOTHESIS TESTING

### 7.1 Try It Yourself Solutions

**1a.** (1) The mean is not 74 months.
$\mu \neq 74$

(2) The variance is less than or equal to 2.7.
$\sigma^2 \leq 2.7$

(3) The proportion is more than 24%
$p > 0.24$

**b.** (1) $\mu = 74$      (2) $\sigma^2 > 2.7$      (3) $p \leq 0.24$

**c.** (1) $H_0 : \mu = 74$; $H_a : \mu \neq 74$ (claim)

(2) $H_0 : \sigma^2 \leq 2.7$ (claim); $H_a : \sigma^2 > 2.7$

(3) $p \leq 0.24$; $H_a : p > 0.24$ (claim)

**2a.** $H_0 : p \leq 0.01$; $H_a : p > 0.01$

**b.** A type I error will occur if the actual proportion is less than or equal to 0.01, but you reject $H_0$.

A type II error will occur if the actual proportion is greater than 0.01, but you fail to reject $H_0$.

**c.** A type II error is more serious because you would be misleading the consumer, possibly causing serious injury or death.

**3a.** (1) $H_0$ : The mean life of a certain type of automobile battery is 74 months.

$H_a$ : The mean life of a certain type of automobile battery is not 74 months.

$H_0 : \mu = 74$; $H_a : \mu \neq 74$

(2) $H_0$ : The proportion of homeowners who feel their house is too small for their family is less than or equal to 24%.

$H_a$ : The proportion of homeowners who feel their house is too small for their family is greater than 24%.

$H_0 : p \leq 0.24$; $H_a : p > 0.24$

**b.** (1) Two-tailed      (2) Right-tailed

**c.** (1)          (2)

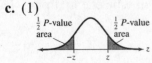 

**4a.** There us enough evidence to support the realtor's claim that the proportion of homeowners who feel their house is too small for their family is more than 24%.

**b.** There is not enough evidence to support the realtor's claim that the proportion of homeowners who feel their house is too small for their family is more than 24%.

**5a.** (1) Support claim.          (2) Reject claim.
  **b.** (1) $H_0 : \mu \geq 650$; $H_a : \mu < 650$ (claim)
      (2) $H_0 : \mu = 98.6$ (claim); $H_a : \mu \neq 98.6$

## 7.1 EXERCISE SOLUTIONS

1.  The two types of hypotheses used in a hypothesis test are the null hypothesis and the alternative hypothesis.
    The alternative hypothesis is the complement of the null hypothesis.

3.  You can reject the null hypothesis, or you can fail to reject the null hypothesis.

5.  False. In a hypothesis test, you assume the null hypothesis is true.

7.  True

9.  False. A small $P$-value in a test will favor a rejection of the null hypothesis.

11. $H_0 : \mu \leq 645$ (claim); $H_a : \mu > 645$          13.          $H_0 : \sigma = 5$; $H_a : \sigma \neq 5$ (claim)

15. $H_0 : p \geq 0.45$; $H_a : p < 0.45$ (claim)

17. c; $H_0 : \mu \leq 3$

19. b; $H_0 : \mu = 3$

21. Right-tailed          **23.** Two-tailed

25. $\mu > 6$                                27. $\sigma \leq 320$
    $H_0 : \mu \leq 6$; $H_a : \mu > 6$ (claim)          $H_0 : \sigma \leq 320$ (claim); $H_a : \sigma > 320$

29. $\mu < 45$
    $H_0 : \mu \geq 45$; $H_a : \mu < 45$ (claim)

31. A type I error will occur if the actual proportion of new customers who return to buy their next piece of furniture is at least 0.60, but you reject $H_0 : p \geq 0.60$.
    A type II error will occur if the actual proportion of new customers who return to buy their next piece of furniture is less than 0.60, but you fail to reject $H_0 : p \geq 0.60$.

33. A type I error will occur if the actual standard deviation of the length of time to play a game is less than or equal to 12 minutes, but you reject $H_0 : \sigma \leq 12$.
    A type II error will occur of the actual standard deviation of the length of time to play a game is greater than 12 minutes, but you fail to reject $H_0 : \sigma \leq 12$.

**35.** A type I error will occur if the actual proportion of applicants who become police officers is at most 0.20, but you reject $H_0 : p \leq 0.20$.

A type II error will occur if the actual proportion of applicants who become police officers is greater than 0.20, but you fail to reject $H_0 : p \leq 0.20$.

**37.** $H_0$ : The proportion of homeowners who have a home security alarm is greater than or equal to 14%.

$H_a$ : The proportion of homeowners who have a home security alarm is less than 14%.

$H_0 : p \geq 0.14$; $H_a : p < 0.14$

Left-tailed because the alternative hypothesis contains <.

**39.** $H_0$ : The standard deviation of the 18-hole scores for a golfer is greater than or equal to 2.1 strokes.

$H_a$ : The standard deviation of the 18-hole scores for a golfer is less than 2.1 strokes.

$H_0 : \sigma \geq 2.1$; $H_a : \sigma < 2.1$

Left-tailed because the alternative hypothesis contains <.

**41.** $H_0$ : The mean length of the baseball team's games is greater than or equal to 2.5 hours.

$H_a$ : The mean length of the baseball team's games is less than 2.5 hours.

$H_0 : \mu \geq 2.5$; $H_a : \mu < 2.5$

Left-tailed because the alternative hypothesis contains <.

**43.** Alternative hypothesis
   (a) There is enough evidence to support the scientist's claim that the mean incubation period for swan eggs is less than 40 days.
   (b) There is not enough evidence to support the scientist's claim that the mean incubation period for swan eggs is less than 40 days.

**45.** Null hypothesis
   (a) There is enough evidence to reject the claim that the standard deviation of the life of the lawn mower is at most 2.8 years.
   (b) There is not enough evidence to reject the claim that the standard deviation of the life of the lawnmower is at most 2.8 years.

**47.** Alternative hypothesis
   (a) There is enough evidence to support the researcher's claim that less than 16% of people had no health care visits in the past year.
   (b) There is not enough evidence to support the researcher's claim that less than 16% of people had no health care visits in the past year.

**49.** $H_0 : \mu \geq 60$; $H_a : \mu < 60$

**51.** (a) $H_0 : \mu \geq 15$; $H_a : \mu < 15$
   (b) $H_0 : \mu \leq 15$; $H_a : \mu > 15$

**53.** If you decrease $\alpha$, you are decreasing the probability that you reject $H_0$. Therefore, you are increasing the probability of failing to reject $H_0$. This could increase $\beta$, the probability of failing to reject $H_0$ when $H_0$ is false.

**55.** Yes; If the $P$-value is less than $\alpha = 0.05$, it is also less than $\alpha = 0.10$.

**57.** (a) Fail to reject $H_0$ because the CI includes values greater than 70.
(b) Reject $H_0$ because the CI is located below 70.
(c) Fail to reject $H_0$ because the CI includes values greater than 70.

**59.** (a) Reject $H_0$ because the CI is located to the right of 0.20.
(b) Fail to reject $H_0$ because the CI includes values less than 0.20.
(c) Fail to reject $H_0$ because the CI includes values less than 0.20.

## 7.2 HYPOTHESIS TESTING FOR THE MEAN (LARGE SAMPLES)

### 7.2 Try It Yourself Solutions

**1a.** (1) $P = 0.0745 > 0.05 = \alpha$
(2) $P = 0.0745 < 0.10 = \alpha$
**b.** (1) Fail to reject $H_0$ because $0.0745 > 0.05$.
(2) Reject $H_0$ because $0.0745 < 0.10$.

**2a.**

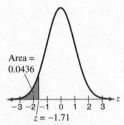

**b.** $P = 0.0436$
**c.** Reject $H_0$ because $P = 0.0436 < 0.05$.

**3a.** Area to the left of $z = 1.64$ is $0.9495$.
**b.** $P = 2(\text{area}) = 2(0.0505) = 0.1010$

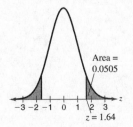

**c.** Fail to reject $H_0$ because $P = 0.1010 > 0.10$.

**4a.** The claim is "the mean speed is greater than 35 miles per hour."
$H_0 : \mu \leq 35$; $H_\alpha : \mu > 35$ (claim)

**b.** $\alpha = 0.05$

**c.** $z = \dfrac{\bar{x} - \mu}{\dfrac{s}{\sqrt{n}}} = \dfrac{36 - 35}{\dfrac{4}{\sqrt{100}}} = 2.5$

**d.** $P$-value = {Area right of $z = 2.50$} $= 0.0062$

**e.** Reject $H_0$ because $P$-value $= 0.0062 < 0.05$.

**f.** There is enough evidence at the 5% level of significance to support the claim that the average speed is greater than 35 miles per hour.

**5a.** The claim is "the mean time to recoup the cost of bariatric surgery is 3 years."
$H_0 : \mu = 3$ (claim); $H_\alpha : \mu \neq 3$

**b.** $\alpha = 0.01$

**c.** $z = \dfrac{\bar{x} - \mu}{\dfrac{s}{\sqrt{n}}} = \dfrac{3.3 - 3}{\dfrac{0.5}{\sqrt{25}}} = 3$

**d.** $P$-value = 2{Area to the right of $z = 3.00$} $= 2(0.0013) = 0.0026$

**e.** Reject $H_0$ because $P$-value $= 0.0026 < 0.01$.

**f.** There is enough evidence at the 1% level of significance to reject the claim that the mean time to recover from bariatric surgery is 3 years.

**6a.** $P = 0.0440 > 0.01 = \alpha$          **b.** Fail to reject $H_0$.

**7a.**

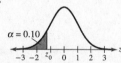

**b.** Area $= 0.1003$

**c.** $z_0 = -1.28$

**d.** Rejection region: $z < -1.28$

**8a.**

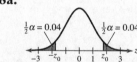

**b.** 0.0401 and 0.9599

**c.** $-z_0 = -1.75$ and $z_0 = 1.75$

**d.** Rejection region: $z < -1.75$, $z > 1.75$

**9a.** The claim is "the mean work day of the company's mechanical engineers is less than 8.5 hours."
$H_0 : \mu \geq 8.5$; $H_a : \mu < 8.5$ (claim)

**b.** $\alpha = 0.01$

**c.** $z_0 = -2.33$; Rejection region: $z < -2.33$

**d.** $z = \dfrac{\bar{x} - \mu}{\dfrac{s}{\sqrt{n}}} = \dfrac{8.2 - 8.5}{\dfrac{0.5}{\sqrt{25}}} = -3.00$

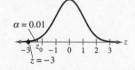

**e.** Because $-3.00 < -2.33$, reject $H_0$.

**f.** There is enough evidence at the 1% level of significance to support the claim that the mean work day is less than 8.5 hours.

**10a.** $\alpha = 0.01$

**b.** $-z_0 = -2.575$, $z_0 = 2.575$; Rejection regions: $z < -2.575$, $z > 2.575$

**c.**

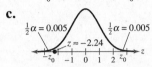

Fail to reject $H_0$.

**d.** There is not enough evidence at the 1% level of significance to reject the claim that the mean cost of raising a child (age 2 and under) by husband-wife families in the United States is $13,960.

## 7.2 EXERCISE SOLUTIONS

**1.** In the $z$-test using rejection region(s), the test statistic is compared with critical values. The $z$-test using a $P$-value compares the $P$-value with the level of significance $\alpha$.

**3.** (a) Fail to reject $H_0$ because $P = 0.0461 > 0.01 = \alpha$.
   (b) Reject $H_0$ because $P = 0.0461 < 0.05 = \alpha$.
   (c) Reject $H_0$ because $P = 0.0461 < 0.10 = \alpha$.

**5.** (a) Fail to reject $H_0$ because $P = 0.1271 > 0.01 = \alpha$.
   (b) Fail to reject $H_0$ because $P = 0.1271 > 0.05 = \alpha$.
   (c) Fail to reject $H_0$ because $P = 0.1271 > 0.10 = \alpha$.

**7.** (a) Fail to reject $H_0$ because $P = 0.0107 > 0.01 = \alpha$.
   (b) Reject $H_0$ because $P = 0.0107 < 0.05 = \alpha$.
   (c) Reject $H_0$ because $P = 0.0107 < 0.10 = \alpha$.

**9.**

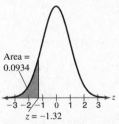

$P = 0.0934$ ; Reject $H_0$ because $P = 0.0934 < 0.10$.

**11.**

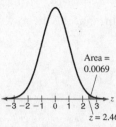

$P = 0.0069$ ; Reject $H_0$ because $P = 0.0069 < 0.01$.

**13.**

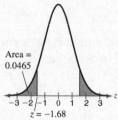

$P = 2(\text{Area}) = 2(0.0465) = 0.0930$ ; Fail to reject $H_0$ because $P = 0.0930 > 0.05$.

**15.** (a) $P = 0.3050$        (b) $P = 0.0089$

The larger $P$-value corresponds to the larger area.

**17.** Fail to reject $H_0$, $(P = 0.0628 > 0.05)$.

**19.** Critical value: $z_0 = -1.88$ ; Rejection region: $z < -1.88$

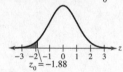

**21.** Critical value: $z_0 = 1.645$ ; Rejection region: $z > 1.645$

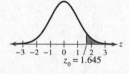

**23.** Critical values: $-z_0 = -2.33$, $z_0 = 2.33$ ; Rejection regions: $z < -2.33$, $z > 2.33$

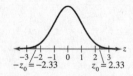

**25.** (a) Fail to reject $H_0$ because $z < 1.285$.

(b) Fail to reject $H_0$ because $z < 1.285$.

(c) Fail to reject $H_0$ because $z < 1.285$.

(d) Reject $H_0$ because $z > 1.285$.

**27.** $H_0 : \mu = 40$ (claim); $H_a : \mu \neq 40$

$\alpha = 0.05 \rightarrow z_0 = \pm 1.96$

$z = \dfrac{\overline{x} - \mu}{\dfrac{s}{\sqrt{n}}} = \dfrac{39.2 - 40}{\dfrac{1.97}{\sqrt{25}}} \approx -2.030$

Because $z = -2.03 < -1.96$, reject $H_0$. There is enough evidence at the 5% level of significance to reject the claim.

**29.** $H_0 : \mu = 8550$; $H_a : \mu \neq 8550$ (claim)

$\alpha = 0.02 \rightarrow z_0 = \pm 2.33$

$z = \dfrac{\overline{x} - \mu}{\dfrac{s}{\sqrt{n}}} = \dfrac{8420 - 8550}{\dfrac{314}{\sqrt{38}}} \approx -2.552$

Because $z = -2.552 < -2.33$, reject $H_0$. There is enough evidence at the 2% level of significance to support the claim.

**31.** (a) The claim is "the mean raw score for the school's applicants is more than 30."

   $H_0 : \mu \leq 30$; $H_a : \mu > 30$ (claim)

   (b) $z = \dfrac{\overline{x} - \mu}{\dfrac{s}{\sqrt{n}}} = \dfrac{31 - 30}{\dfrac{2.5}{\sqrt{50}}} \approx 2.83$

   Area = 0.9977

   (c) $P$-value = {Area to right of $z = 2.83$ } = 0.0023

   (d) Because $P = 0.0023 < 0.01 = \alpha$, reject $H_0$.

   (e) There is enough evidence at the 1% level of significance to support the claim that the mean raw score for the school's applicants is more than 30.

**33.** (a) The claim is "the mean annual consumption of cheddar cheese by a person in the United States is at most 10.3 pounds."

   $H_0 : \mu \leq 10.3$ (claim); $H_a : \mu > 10.3$

   (b) $z = \dfrac{\overline{x} - \mu}{\dfrac{s}{\sqrt{n}}} = \dfrac{9.9 - 10.3}{\dfrac{2.1}{\sqrt{100}}} \approx -1.90$

   Area = 0.0287

   (c) $P$-value = {Area to the right of $z = -1.90$ } = 0.9713

   (d) Because $P = 0.9713 > 0.05 = \alpha$, fail to reject $H_0$.

   (e) There is not enough evidence at the 5% level of significance to reject the claim that the mean annual consumption of cheddar cheese by a person in the United States is at most 10.3 pounds.

**35.** (a) The claim is "the mean time it takes smokers to quit smoking permanently is 15 years."

   $H_0 : \mu = 15$ (claim); $H_a : \mu \neq 15$

(b) $\bar{x} \approx 14.834$

$$z = \frac{\bar{x} - \mu}{\frac{\sigma}{\sqrt{n}}} = \frac{14.834 - 15}{\frac{6.2}{\sqrt{32}}} \approx -0.15$$

Area = 0.4404

(c) $P$-value = 2{Area to left of $z = -0.15$ } = 2(0.4404) = 0.8808

(d) Because $P = 0.8808 > 0.05 = \alpha$, fail to reject $H_0$.

(e) There is not enough evidence at the 5% level of significance to reject the claim that the mean time it takes smokers to quit smoking permanently is 15 years.

37. (a) The claim is "the mean caffeine content per 12-ounce bottle of cola is 40 milligrams."
    $H_0$ : $\mu = 40$ (claim); $H_a$ : $\mu \neq 40$

(b) $-z_0 = -2.575$, $z_0 = 2.575$;
    Rejection regions: $z < -2.575$, $z > 2.575$

(c) $z = \frac{\bar{x} - \mu}{\frac{s}{\sqrt{n}}} = \frac{39.2 - 40}{\frac{7.5}{\sqrt{20}}} \approx -0.477$

(d) Because $-2.575 < z < 2.575$, fail to reject $H_0$.

(e) There is not enough evidence at the 1% level of significance to reject the claim that the mean caffeine content per 12-ounce bottle of cola is 40 milligrams.

39. (a) The claim is "the mean sodium content in one of its breakfast sandwiches is no more than 920 milligrams."
    $H_0$ : $\mu \leq 920$ (claim); $H_a$ : $\mu > 920$

(b) $z_0 = 1.28$; Rejection region: $z > 1.28$

(c) $z = \frac{\bar{x} - \mu}{\frac{\sigma}{\sqrt{n}}} = \frac{925 - 920}{\frac{18}{\sqrt{44}}} \approx 1.84$

(d) Because $z > 1.28$, reject $H_0$.

(e) There is enough evidence at the 10% level of significance to reject the claim that the mean sodium content in one of their breakfast sandwiches is no more than 920 milligrams.

41. (a) The claim is "the mean nitrogen dioxide level in Calgary is greater than 32 parts per billion."
    $H_0$ : $\mu \leq 32$; $H_a$ : $\mu > 32$ (claim)

(b) $z_0 = 1.555$; Rejection region: $z > 1.555$

(c) $\bar{x} \approx 29.7$

$$z = \frac{\bar{x} - \mu}{\frac{s}{\sqrt{n}}} \approx \frac{29.7 - 32}{\frac{9}{\sqrt{34}}} \approx -1.49$$

(d) Because $z < 1.555$, fail to reject $H_0$.

(e) There is not enough evidence at the 6% level of significance to support the claim that the mean nitrogen dioxide level in Calgary is greater than 32 parts per billion.

43. Outside; When the standardized test statistic is inside the rejection region, $P < \alpha$.

## 7.3 HYPOTHESIS TESTING FOR THE MEAN (SMALL SAMPLES)

### 7.3 Try It Yourself Solutions

**1a.** 13      **b.** $t_0 = -2.650$

**2a.** 8      **b.** $t_0 = 1.397$

**3a.** 15      **b.** $-t_0 = -2.131$, $t_0 = 2.131$

**4a.** The claim is "the mean cost of insuring a two-year-old sedan (in good condition) is less than $1200."
$H_0 : \mu \geq \$1200$; $H_a : \mu < \$1200$ (claim)
**b.** $\alpha = 0.10$ and d.f. $= n - 1 = 6$
**c.** $t_0 = -1.440$; Rejection region: $t < -1.440$
**d.** $t = \dfrac{\bar{x} - \mu}{\dfrac{s}{\sqrt{n}}} = \dfrac{1125 - 1200}{\dfrac{55}{\sqrt{7}}} \approx -3.61$

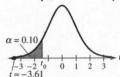

**e.** Because $t < -1.440$, reject $H_0$.
**f.** There is enough evidence at the 10% level of significance to support the claim that the mean cost of insuring a two-year-old sedan (in good condition) is less than $1200.

**5a.** The claim is "the mean conductivity of the river is 1890 milligrams per liter."
$H_0 : \mu = 1890$ (claim); $H_a : \mu \neq 1890$
**b.** $\alpha = 0.01$ and d.f. $= n - 1 = 38$
**c.** $-t_0 = -2.712$, $t_0 = 2.712$; Rejection regions: $t < -2.712$, $t > 2.712$
**d.** $t = \dfrac{\bar{x} - \mu}{\dfrac{s}{\sqrt{n}}} = \dfrac{2350 - 1890}{\dfrac{900}{\sqrt{39}}} \approx 3.192$

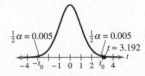

**e.** Because $t > 2.712$, reject $H_0$.
**f.** There is enough evidence at the 1% level of significance to reject the claim that the mean conductivity of the river is 1890 milligrams per liter.

**6a.** The claim is "the mean wait time is at most 18 minutes."
$H_0 : \mu \leq 18$ minutes (claim); $H_a : \mu > 18$ minutes
**b.** $P-\text{value} = 0.9997$

**c.** $P - \text{value} = 0.9997 > 0.05 = \alpha$
Fail to reject $H_0$.

**d.** There is not enough evidence at the 5% level of significance to reject the claim that the mean wait time is at most 18 minutes.

## 7.3 EXERCISE SOLUTIONS

1. Identify the level of significance $\alpha$ and the degrees of freedom, d.f. $= n - 1$. Find the critical value(s) using the $t$-distribution table in the row with $n - 1$ d.f. If the hypothesis test is:
   (1) left-tailed, use the "One Tail, $\alpha$" column with a negative sign.
   (2) right-tailed, use the "One Tail, $\alpha$" column with a positive sign.
   (3) two-tailed, use the "Two Tail, $\alpha$," column with a negative and a positive sign.

3. Critical value: $t_0 = -1.328$; Rejection region: $t < -1.328$

5. Critical value: $t_0 = 1.717$; Rejection region: $\qquad t > 1.717$

7. Critical values: $t_0 = -2.056, t_0 = 2.056$; Rejection regions: $t < -2.056, t > 2.056$

9. (a) Fail to reject $H_0$ because $t > -2.086$.
   (b) Fail to reject $H_0$ because $t > -2.086$.
   (c) Fail to reject $H_0$ because $t > -2.086$.
   (d) Reject $H_0$ because $t < -2.086$.

11. $H_0: \mu = 15$ (claim); $H_a: \mu \neq 15$
    $\alpha = 0.01$ and d.f. $= n - 1 = 35$
    $t_0 = \pm 2.724$
    $$t = \frac{\bar{x} - \mu}{\dfrac{s}{\sqrt{n}}} = \frac{13.9 - 15}{\dfrac{3.23}{\sqrt{36}}} \approx -2.043$$
    Because $-2.724 < t < 2.724$, fail to reject $H_0$. There is not enough evidence at the 1% level of significance to reject the claim.

13. $H_0: \mu \geq 8000$ (claim); $H_a: \mu < 8000$
    $\alpha = 0.01$ and d.f. $= n - 1 = 24$
    $t_0 = -2.492$
    $$t = \frac{\bar{x} - \mu}{\dfrac{s}{\sqrt{n}}} = \frac{7700 - 8000}{\dfrac{450}{\sqrt{25}}} \approx -3.333$$
    Because $t < -2.492$, reject $H_0$. There is enough evidence at the 1% level of significance to reject the claim.

15. (a) The claim is "the mean price of a three-year-old sports utility vehicle (in good condition) is $20,000."

$H_0 : \mu = 20,000$ (claim); $H_a : \mu \neq 20,000$

(b) $-t_0 = -2.080$, $t_0 = 2.080$; Rejection region: $t < -2.080$, $t > 2.080$

(c) $t = \dfrac{\bar{x} - \mu}{\frac{s}{\sqrt{n}}} = \dfrac{20,640 - 20,000}{\frac{1990}{\sqrt{22}}} \approx 1.508$

(d) Because $-2.080 < t < 2.080$, fail to reject $H_0$.

(e) There is not enough evidence at the 5% level of significance to reject the claim that the mean price of a three-year-old sports utility vehicle (in good condition) is $20,000.

17. (a) The claim is "the mean credit card debt for individuals is greater than $5000."

$H_0 : \mu \leq 5000$ ; $H_a : \mu > 5000$ (claim)

(b) $t_0 = 1.688$; Rejection region: $t > 1.688$

(c) $t = \dfrac{\bar{x} - \mu}{\frac{s}{\sqrt{n}}} = \dfrac{5122 - 5000}{\frac{625}{\sqrt{37}}} \approx 1.19$

(d) Because $t < 1.688$, fail to reject $H_0$.

(e) There is not enough evidence at the 5% level of significance to support the claim that the mean credit card debt for individuals is greater than $5000.

19. (a) The claim is "the mean amount of waste recycled by adults in the United States is more than 1 pound per person per day."

$H_0 : \mu \leq 1$; $H_a : \mu > 1$ (claim)

(b) $t_0 = 1.356$; Rejection region: $t > 1.356$

(c) $t = \dfrac{\bar{x} - \mu}{\frac{s}{\sqrt{n}}} = \dfrac{1.51 - 1}{\frac{0.28}{\sqrt{13}}} \approx 6.57$

(d) Because $t > 1.356$, reject $H_0$.

(e) There is enough evidence at the 10% level of significance to support the claim that the mean amount of waste recycled by adults in the United States is more than 1 pound per person per day.

21. (a) The claim is "the mean annual salary for full-time male workers over age 25 and without a high school diploma is $26,000."

$H_0 : \mu = \$26,000$ (claim); $H_a : \mu \neq \$26,000$

(b) $-t_0 = -2.262$, $t_0 = 2.262$; Rejection region: $t < -2.262$, $t > 2.262$

(c) $\bar{x} \approx 25,352.2$, $\qquad\qquad s \approx \$3197.1$

$t = \dfrac{\bar{x} - \mu}{\frac{s}{\sqrt{n}}} \approx \dfrac{25,352.2 - 26,000}{\frac{3197.1}{\sqrt{10}}} \approx -0.64$

(d) Because $-2.262 < t < 2.262$, fail to reject $H_0$.

(e) There is not enough evidence at the 5% level of significance to reject the claim that the mean salary for full-time male workers over age 25 without a high school diploma is $26,000.

**23.** (a) The claim is "the mean speed of vehicles is greater than 45 miles per hour."

   $H_0 : \mu \leq 45; H_a : \mu > 45$ (claim)

   (b) $t = \dfrac{\overline{x} - \mu}{\dfrac{s}{\sqrt{n}}} = \dfrac{48 - 45}{\dfrac{5.4}{\sqrt{25}}} \approx 2.78$

   P-value = {Area to right of $t = 2.78$ } = 0.0052

   (c) Because $P < 0.10 = \alpha$, reject $H_0$.

   (d) There is enough evidence at the 10% level of significance to support the claim that the mean speed of the vehicles is greater than 45 miles per hour.

**25.** (a) The claim is "the mean dive depth of a North Atlantic right whale is 115 meters."

   $H_0 : \mu = 115$ (claim); $H_a : \mu \neq 115$

   (b) $t = \dfrac{\overline{x} - \mu}{\dfrac{s}{\sqrt{n}}} = \dfrac{121.2 - 115}{\dfrac{24.2}{\sqrt{34}}} \approx 1.49$

   P-value = 2{Area to right of $t = 1.49$ } = 2(0.07235) = 0.1447

   (c) Because $P > 0.10 = \alpha$, fail to reject $H_0$.

   (d) There is not enough evidence at the 10% level of significance to reject the claim that the mean dive depth of a North Atlantic right whale is 115 meters.

**27.** (a) The claim is "the mean class size for full-time faculty is fewer than 32 students."

   $H_0 : \mu \geq 32; H_a : \mu < 32$ (claim)

   (b) $\overline{x} \approx 30.167$           $s \approx 4.004$

   $t = \dfrac{\overline{x} - \mu}{\dfrac{s}{\sqrt{n}}} = \dfrac{30.167 - 32}{\dfrac{4.004}{\sqrt{18}}} \approx -1.942$

   P-value = {Area to left of $t = -1.942$ } = 0.0344

   (c) Because $P < 0.05 = \alpha$, reject $H_0$.

   (d) There is enough evidence at the 5% level of significance to support the claim that the mean class size for full-time faculty is fewer than 32 students.

**29.** Because $\sigma$ is unknown, $n < 30$, the sample is random, and the gas mileage is normally distributed, use the $t$-distribution.

   $H_0 : \mu \geq 23$ (claim); $H_a : \mu < 23$

   $t = \dfrac{\overline{x} - \mu}{\dfrac{s}{\sqrt{n}}} \approx \dfrac{22 - 23}{\dfrac{4}{\sqrt{5}}} \approx -0.559$

   P-value = {Area left of $t = t = -0.559$ } = 0.3030

   Because $P > 0.05 = \alpha$, fail to reject $H_0$. There is not enough evidence at the 5% level of significance to reject the claim that the mean gas mileage for the luxury sedan is at least 23 miles per gallon.

**31.** More likely; The tails of the $t$-distribution curve are thicker than those of a standard normal distribution curve. So, if you incorrectly use a standard normal sampling distribution instead of a $t$-sampling distribution, the area under the curve at the tails will be smaller than what it would be for the $t$-test, meaning the critical value(s) will lie closer to the mean. This makes it more likely for the test statistic to be in the rejection region(s). This result is the same regardless of whether the test is left-tailed, right-tailed, or two-tailed; in each case, the tail thickness affects the location of the critical value(s).

## 7.4 HYPOTHESIS TESTING FOR PROPORTIONS

### 7.4 Try It Yourself Solutions

**1a.** $np = (150)(0.30) = 45 > 5$, $nq = (150)(0.70) = 105 > 5$

**b.** The claim is "more than 30% of U.S. smartphone owners use their phones while watching television."
$$H_0 : p \leq 0.30; \ H_a : p > 0.30 \ \text{(claim)}$$

**c.** $\alpha = 0.05$

**d.** $z_0 = 1.645$; Rejection region: $z > 1.645$

**e.** $z = \dfrac{\hat{p} - p}{\sqrt{\dfrac{pq}{n}}} = \dfrac{0.38 - 0.30}{\sqrt{\dfrac{(0.30)(0.70)}{150}}} \approx 2.14$

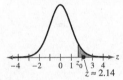

**f.** Reject $H_0$.

**g.** There is enough evidence at the 5% level of significance to support the claim that more than 30% of U.S. smartphone owners use their phones while watching television.

**2a.** $np = (250)(0.30) = 75 > 5$, $nq = (250)(0.70) = 175 > 5$

**b.** The claim is "30% of U.S. adults have not purchased a certain brand because they found the advertisements distasteful."
$$H_0 : p = 0.30 \ \text{(claim)}; \ H_a : p \neq 0.30$$

**c.** $\alpha = 0.10$

**d.** $-z_0 = -1.645$, $z_0 = 1.645$; Rejection region: $z < -1.645$, $z > 1.645$

**e.** $z = \dfrac{\hat{p} - p}{\sqrt{\dfrac{pq}{n}}} = \dfrac{0.36 - 0.30}{\sqrt{\dfrac{(0.30)(0.70)}{250}}} \approx 2.07$

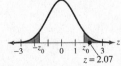

**f.** Reject $H_0$.

**g.** There is enough evidence at the 10% level of significance to reject the claim that 30% of U.S. adults have not purchased a certain brand because they found the advertisements distasteful.

## 7.4 EXERCISE SOLUTIONS

1. If $np \geq 5$ and $nq \geq 5$, the normal distribution can be used.

3. $np = (40)(0.12) = 4.8 < 5$

   $nq = (40)(0.88) = 35.2 > 5$

   Cannot use normal distribution because $np < 5$.

5. $np = (500)(0.15) = 75 > 5$

   $nq = (500)(0.85) = 425 > 5 \rightarrow$ use normal distribution

   $H_0: p = 0.15$; $H_a: p \neq 0.15$ (claim)

   $-z_0 = -1.96$, $z_0 = 1.96$; Rejection region: $z < -1.96$, $z > 1.96$

   $$z = \frac{\hat{p} - p}{\sqrt{\frac{pq}{n}}} = \frac{0.12 - 0.15}{\sqrt{\frac{(0.15)(0.85)}{500}}} \approx -1.88$$

   Fail to reject $H_0$. There is not enough evidence at the 5% level of significance to support the claim.

7. $np = (100)(0.45) = 45 > 5$

   $nq = (100)(0.55) = 55 > 5 \rightarrow$ use normal distribution

   $H_0: p \leq 0.45$ (claim); $H_a: p > 0.45$

   $z_0 = 1.645$; Rejection region: $z > 1.645$

   $$z = \frac{\hat{p} - p}{\sqrt{\frac{pq}{n}}} = \frac{0.52 - 0.45}{\sqrt{\frac{(0.45)(0.55)}{100}}} \approx 1.41$$

   Fail to reject $H_0$. There is not enough evidence at the 5% level of significance to reject the claim.

9. (a) The claim is "less than 25% of U.S. adults are smokers."

   $H_0: p \geq 0.25$; $H_a: p < 0.25$ (claim)

   (b) $z_0 = -1.645$; Rejection region: $z < -1.645$

   (c) $$z = \frac{\hat{p} - p}{\sqrt{\frac{pq}{n}}} = \frac{0.193 - 0.25}{\sqrt{\frac{(0.25)(0.75)}{200}}} \approx -1.86$$

   (d) Because $z < -1.645$, reject $H_0$.

   (e) There is enough evidence at the 5% level of significance to support the claim that less than 25% of U.S. adults are smokers.

11. (a) The claim is "at most 75% of U.S. adults think that drivers are safer using hands-free cell phones instead of using hand-held cell phones."

    $H_0: p \leq 0.75$ (claim); $H_a: p > 0.75$

(b) $z_0 = 2.33$; Rejection region: $z > 2.33$

(c) $z = \dfrac{\hat{p} - p}{\sqrt{\dfrac{pq}{n}}} = \dfrac{0.77 - 0.75}{\sqrt{\dfrac{(0.75)(0.25)}{150}}} \approx 0.57$

(d) Because $z < 2.33$, fail to reject $H_0$.

(e) There is not enough evidence at the 1% level of significance to reject the claim that at most 75% of U.S. adults think that drivers are safer using hands-free cell phones instead of using hand-held cell phones.

13. (a) The claim is "more than 80% of females ages 20-29 are taller than 62 inches."
$H_0 : p \le 0.80$; $H_a : p > 0.80$ (claim)

(b) $z_0 = 1.28$; Rejection region: $z > 1.28$

(c) $z = \dfrac{\hat{p} - p}{\sqrt{\dfrac{pq}{n}}} = \dfrac{0.79 - 0.80}{\sqrt{\dfrac{(0.80)(0.20)}{150}}} \approx -0.31$

(d) Because $z < 1.28$, fail to reject $H_0$.

(e) There is not enough evidence at the 10% level of significance to support the claim that more than 80% of females ages 20-29 are taller than 62 inches.

15. (a) The claim is "less than 35% of U.S. households own a dog."
$H_0 : p \ge 0.35$; $H_a : p < 0.35$ (claim)

(b) $z_0 = -1.28$; Rejection region: $z < -1.28$

(c) $z = \dfrac{\hat{p} - p}{\sqrt{\dfrac{pq}{n}}} = \dfrac{0.39 - 0.35}{\sqrt{\dfrac{(0.35)(0.65)}{400}}} \approx 1.68$

(d) Because $z > -1.28$, fail to reject $H_0$.

(e) There is not enough evidence at the 10% level of significance to support the claim that less than 35% of U.S. households own a dog.

17. $H_0 : p \ge 0.52$ (claim); $H_a : p < 0.52$
$z_0 = -1.645$: Rejection region: $z < -1.645$

$z = \dfrac{\hat{p} - p}{\sqrt{\dfrac{pq}{n}}} = \dfrac{0.48 - 0.52}{\sqrt{\dfrac{(0.52)(0.48)}{50}}} \approx -0.566$

Because $z > -1.645$, fail to reject $H_0$. There is not enough evidence at the 5% level of significance to reject the claim that at least 52% of adults are more likely to buy a product when there are free samples.

19. (a) The claim is "less than 35% of U.S. households own a dog."
$H_0 : p \ge 0.35$; $H_a : p < 0.35$ (claim)

(b) $z_0 = -1.28$: Rejection region: $z < -1.28$

(c) $z = \dfrac{x - np}{\sqrt{npq}} = \dfrac{156 - (400)(0.35)}{\sqrt{(400)(0.35)(0.65)}} \approx 1.68$

(d) Because $z > -1.28$, fail to reject $H_0$.

(e) There is not enough evidence at the 10% level of significance to support the claim that less than 35% of U.S. households own a dog. The results are the same.

## 7.5 HYPOTHESIS TESTING FOR VARIANCE AND STANDARD DEVIATION

## 7.5 Try It Yourself Solutions

**1a.** d.f. $= 17$, $\alpha = 0.01$

**b.** $\chi_0^2 = 33.409$

**2a.** d.f. $= 29$, $\alpha = 0.05$

**b.** $\chi_0^2 = 17.708$

**3a.** d.f. $= 50$, $\alpha = 0.01$

**b.** $\chi_R^2 = 79.490$

**c.** $\chi_L^2 = 27.991$

**4a.** The claim is "the variance of the amount of sports drink in a 12-ounce bottle is no more than 0.40."

$H_0 : \sigma^2 \le 0.40$ (claim); $H_a : \sigma^2 > 0.40$

**b.** $\alpha = 0.01$ and d.f. $= n - 1 = 30$

**c.** $\chi_0^2 = 50.892$; Rejection region: $\chi^2 > 50.892$

**d.** $\chi^2 = \dfrac{(n-1)s^2}{\sigma^2} = \dfrac{(30)(0.75)}{0.40} = 56.250$

**e.** Because $\chi^2 > 50.892$, reject $H_0$.

**f.** There is enough evidence at the 1% level of significance to reject the claim that the variance of the amount of sports drink in a 12-ounce bottle is no more than 0.40.

**5a.** The claim is "the standard deviation of the lengths of response times is less than 3.7 minutes."

$H_0 : \sigma \ge 3.7$; $H_a : \sigma < 3.7$ (claim)

**b.** $\alpha = 0.05$ and d.f. $= n - 1 = 8$

**c.** $\chi_0^2 = 2.733$; Rejection region: $\chi^2 < 2.733$

**d.** $\chi^2 = \dfrac{(n-1)s^2}{\sigma^2} = \dfrac{(8)(3.0)^2}{(3.7)^2} \approx 5.259$

**e.** Because $\chi^2 > 2.733$, fail to reject $H_0$.

**f.** There is not enough evidence at the 5% level of significance to support the claim that the standard deviation of the lengths of response times is less than 3.7 minutes.

**6a.** The claim is "the variance of the weight losses is 25.5."

$H_0 : \sigma^2 = 25.5$ (claim); $H_a : \sigma^2 \neq 25.5$

**b.** $\alpha = 0.10$ and d.f. $= n - 1 = 12$

**c.** $\chi_L^2 = 5.226$ and $\chi_R^2 = 21.026$; Rejection region: $\chi^2 > 21.026$, $\chi^2 < 5.226$

**d.** $\chi^2 = \dfrac{(n-1)s^2}{\sigma^2} = \dfrac{(12)(10.8)}{25.5} \approx 5.082$

**e.** Because $\chi^2 < 5.226$, reject $H_0$.

**f.** There is enough evidence at the 10% level of significance to reject the claim that the variance of the weight losses of users is 25.5.

## 7.5 EXERCISE SOLUTIONS

**1.** Specify the level of significance $\alpha$. Determine the degrees of freedom. Determine the critical values using the $\chi^2$-distribution. For a right-tailed test, use the value that corresponds to d.f. and $\alpha$. For a left-tailed test, use the value that corresponds to d.f. and $1 - \alpha$. For a two-tailed test, use the value that corresponds to d.f. and $\dfrac{1}{2}\alpha$, and d.f. and $1 - \dfrac{1}{2}\alpha$.

**3.** The requirement of a normal distribution is more important when testing a standard deviation than when testing a mean. When the population is not normal, the results of the chi-square test can be misleading because the chi-square test is not as robust as the tests for the population mean.

**5.** Critical value: $\chi_0^2 = 38.885$; Rejection region: $\chi^2 > 38.885$

**7.** Critical value: $\chi_0^2 = 0.872$; Rejection region: $\chi^2 < 0.872$

**9.** Critical values: $\chi_L^2 = 60.391$, $\chi_R^2 = 101.879$; Rejection regions: $\chi^2 < 60.391$, $\chi^2 > 101.879$

**11.** (a) Fail to reject $H_0$ because $\chi^2 < 6.251$.

(b) Fail to reject $H_0$ because $\chi^2 < 6.251$.

(c) Fail to reject $H_0$ because $\chi^2 < 6.251$.

(d) Reject $H_0$ because $\chi^2 > 6.251$.

**13.** $H_0 : \sigma^2 = 0.52$ (claim); $H_a : \sigma^2 \neq 0.52$

$\chi_L^2 = 7.564$, $\chi_R^2 = 30.191$; Rejection regions: $\chi^2 < 7.564$, $\chi^2 > 30.191$

$\chi^2 = \dfrac{(n-1)s^2}{\sigma^2} = \dfrac{(17)(0.508)^2}{(0.52)} \approx 16.608$

Because $7.564 < \chi^2 < 30.191$, fail to reject $H_0$. There is not enough evidence at the 5% level of significance to reject the claim.

**15.** $H_0$: $\sigma = 24.9$ (claim); $H_a$: $\sigma \neq 24.9$

$\chi_L^2 = 34.764$, $\chi_R^2 = 67.505$; Rejection regions: $\chi^2 < 34.764$, $\chi^2 > 67.505$

$$\chi^2 = \frac{(n-1)s^2}{\sigma^2} = \frac{(50)(29.1)^2}{(24.9)^2} \approx 68.29$$

Because $\chi^2 > 67.505$, Reject $H_0$. There is enough evidence at the 10% level of significance to reject the claim.

**17.** (a) The claim is "the variance of the diameters in a certain tire model is 8.6."

    $H_0$: $\sigma^2 = 8.6$ (claim); $H_a$: $\sigma^2 \neq 8.6$

(b) $\chi_L^2 = 1.735$, $\chi_R^2 = 23.589$; Rejection regions: $\chi^2 < 1.735$, $\chi^2 > 23.589$

(c) $\chi^2 = \dfrac{(n-1)s^2}{\sigma^2} = \dfrac{(9)(4.3)}{8.6} = 4.5$

(d) Because $1.735 < \chi^2 < 23.589$, fail to reject $H_0$.

(e) There is not enough evidence at the 1% level of significance to reject the claim that the variance of the diameters in a certain tire model is 8.6.

**19.** (a) The claim is "the standard deviation for eighth-grade students on a science assessment test is less than 36 points."

    $H_0$: $\sigma \geq 36$; $H_a$: $\sigma < 36$ (claim)

(b) $\chi_0^2 = 13.240$; Rejection region: $\chi^2 < 13.240$

(c) $\chi^2 = \dfrac{(n-1)s^2}{\sigma^2} = \dfrac{(21)(33.4)^2}{(36)^2} \approx 18.076$

(d) Because $\chi^2 > 13.240$, fail to reject $H_0$.

(e) There is not enough evidence at the 10% level of significance to support the claim that the standard deviation for eighth-graders on the examination is less than 36 points.

**21.** (a) The claim is "the standard deviation of waiting times experienced by patients is no more than 0.5 minute."

    $H_0$: $\sigma \leq 0.5$ (claim); $H_a$: $\sigma > 0.5$

(b) $\chi_0^2 = 33.196$; Rejection region: $\chi^2 > 33.196$

(c) $\chi^2 = \dfrac{(n-1)s^2}{\sigma^2} = \dfrac{(24)(0.7)^2}{(0.5)^2} \approx 47.04$

(d) Because $\chi^2 > 33.196$, reject $H_0$.

(e) There is enough evidence at the 10% level of significance to reject the claim that the standard deviation of waiting times experienced by patients is no more than 0.5 minute.

**23.** (a) The claim is "the standard deviation of the annual salaries is different from \$5500."

    $H_0$: $\sigma = \$5500$; $H_a$: $\sigma \neq \$5500$ (claim)

(b) $\chi_L^2 = 5.009$, $\chi_R^2 = 24.736$ Rejection regions: $\chi^2 < 5.009$, $\chi^2 > 24.736$

(c) $s \approx 7780.146$

$$\chi^2 = \frac{(n-1)s^2}{\sigma^2} = \frac{(13)(7780.146)^2}{(5500)^2} \approx 26.01$$

(d) Because $\chi^2 > 24.736$, reject $H_0$.

(e) There is enough evidence at the 5% level of significance to support the claim that the standard deviation of the annual salaries is different from \$5500.

25. $\chi^2 = 18.076$, d.f. $= n - 1 = 21$

$P$-value = {Area left of $\chi^2 = 18.076$} $= 0.3558$

Fail to reject $H_0$ because $P$-value $= 0.3558 > 0.10$.

27. $\chi^2 = 47.04$, d.f. $= n - 1 = 24$

$P$-value = {Area right of $\chi^2 = 47.04$} $= 0.0033$

Reject $H_0$ because $P$-value $= 0.0033 < 0.10$.

# CHAPTER 7 REVIEW EXERCISE SOLUTIONS

1. $H_0: \mu \leq 375$ (claim); $H_a: \mu > 375$

3. $H_0: p \geq 0.205$; $H_a: p < 0.205$ (claim)

5. $H_0: \sigma \leq 1.9$; $H_a: \sigma > 1.9$ (claim)

7. (a) $H_0: p = 0.41$ (claim); $H_a: p \neq 0.41$

(b) A type I error will occur if the actual proportion of U.S. adults who say Earth Day has helped raise environmental awareness is 41%, but you reject $H_0: p = 0.41$.

A type II error will occur if the actual proportion is not 41%, but you fail to reject $H_0: p = 0.41$.

(c) Two-tailed because the alternative hypothesis contains $\neq$.

(d) There is enough evidence to reject the news outlet's claim that the proportion of U.S. adults who say Earth Day has helped raise environmental awareness is 41%.

(e) There is not enough evidence to reject the news outlet's claim that the proportion of U.S. adults who say Earth Day has helped raise environmental awareness is 41%.

9. (a) $H_0: \sigma \leq 50$ (claim); $H_a: \sigma > 50$

(b) A type I error will occur if the actual standard deviation of the sodium content in one serving of a certain soup is no more than 50 milligrams, but you reject $H_0: \sigma \leq 50$.

A type II error will occur if the actual standard deviation of the sodium content in one serving of a certain soup is more than 50 milligrams, but you fail to reject $H_0: \sigma \leq 50$.

(c) Right-tailed because the alternative hypothesis contains $>$.

(d) There is enough evidence to reject the soup maker's claim that the standard deviation of the sodium content in one serving of a certain soup is no more than 50 milligrams.

(e) There is not enough evidence to reject the soup maker's claim that the standard deviation of the sodium content in one serving of a certain soup is no more than 50 milligrams.

11. $P$-value = {Area to left of $z = -0.94$} $= 0.1736$

Fail to reject $H_0$.

13. (a) The claim is "the mean annual consumption of coffee by a person in the United States is 23.2 gallons."

$H_0: \mu = 23.2$ (claim); $H_a: \mu \neq 23.2$

(b) $z = \dfrac{\bar{x} - \mu}{\dfrac{\sigma}{\sqrt{n}}} = \dfrac{21.6 - 23.2}{\dfrac{4.8}{\sqrt{90}}} \approx -3.16$

(c) $P$-value $= 2\{$Area to left of $z = -3.16\} = 2(0.0008) = 0.0016$

(d) Because $P$-value $< 0.05 = \alpha$, reject $H_o$.

(e) There is enough evidence at the 5% level of significance to reject the claim that the mean annual consumption of coffee by a person in the United States is 23.2 gallons.

15. Critical value: $z_0 \approx -2.05$; Rejection region: $z < -2.05$

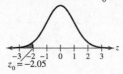

17. Critical value: $z_0 = 1.96$; Rejection region: $z > 1.96$

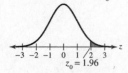

19. Fail to reject $H_0$ because $-1.645 < z < 1.645$.

21. Fail to reject $H_0$ because $-1.645 < z < 1.645$.

23. $H_0: \mu \leq 45$ (claim); $H_a: \mu > 45$

$z_0 = 1.645$; Rejection region: $z > 1.645$

$z = \dfrac{\bar{x} - \mu}{\dfrac{\sigma}{\sqrt{n}}} = \dfrac{47.2 - 45}{\dfrac{6.7}{\sqrt{22}}} \approx 1.54$

Because $z < 1.645$, fail to reject $H_0$. There is not enough evidence at the 5% level of significance to reject the claim.

25. $H_0: \mu \geq 5.500$; $H_a: \mu < 5.500$ (claim)

$z_0 = -2.33$; Rejection region: $z < -2.33$

$z = \dfrac{\bar{x} - \mu}{\dfrac{\sigma}{\sqrt{n}}} = \dfrac{5.497 - 5.500}{\dfrac{0.011}{\sqrt{36}}} \approx -1.636$

Because $z > -2.33$, fail to reject $H_0$. There is not enough evidence at the 1% level of significance to support the claim.

27. (a) The claim is "the mean annual cost of raising a child (age 2 and under) by husband-wife families in rural areas is \$11,060."

$H_0: \mu = \$11,060$ (claim); $H_a: \mu \neq \$11,060$

(b) $-z_0 = -2.575$, $z_0 = 2.575$; Rejection regions: $z < -2.575$, $z > 2.575$

(c) $z = \dfrac{\bar{x} - \mu}{\frac{s}{\sqrt{n}}} = \dfrac{10{,}920 - 11{,}060}{\frac{1561}{\sqrt{800}}} \approx -2.54$

(d) Because $-2.575 < z < 2.575$, fail to reject $H_0$.

(e) There is not enough evidence at the 1% level of significance to reject the claim that the mean cost of raising a child (age 2 and under) by husband-wife families in rural areas is \$11,060.

**29.** Critical values: $-t_0 = -2.093$, $t_0 = 2.093$; Rejection region: $-t < -2.093$, $t > 2.093$

**31.** Critical value: $t_0 = -2.977$; Rejection region: $t < -2.977$

**33.** $H_0: \mu \le 12{,}700$; $H_a: \mu > 12{,}700$ (claim)

$t_0 = 2.845$; Rejection region: $t > 2.845$

$t = \dfrac{\bar{x} - \mu}{\frac{s}{\sqrt{n}}} = \dfrac{12{,}855 - 12{,}700}{\frac{248}{\sqrt{21}}} \approx 2.864$

Because $t > 2.845$, reject $H_0$. There is enough evidence at the 0.5% level of significance to support the claim.

**35.** $H_0: \mu \le 51$ (claim); $H_a: \mu > 51$

$t_0 = 2.426$; Rejection region: $t > 2.426$

$t = \dfrac{\bar{x} - \mu}{\frac{s}{\sqrt{n}}} = \dfrac{52 - 51}{\frac{2.5}{\sqrt{40}}} \approx 2.530$

Because $t > 2.426$, reject $H_0$. There is enough evidence at the 1% level of significance to reject the claim.

**37.** (a) The claim is "the mean monthly cost of joining a health club is \$25."

$H_0: \mu = \$25$ (claim); $H_a: \mu \ne \$25$

(b) $-t_0 = -1.740$, $t_0 = 1.740$; Rejection regions: $t < -1.740$, $t > 1.740$

(c) $t = \dfrac{\bar{x} - \mu}{\frac{s}{\sqrt{n}}} = \dfrac{26.25 - 25}{\frac{3.23}{\sqrt{18}}} \approx 1.642$

(d) Because $-1.740 < t < 1.740$, fail to reject $H_0$.

(e) There is not enough evidence at the 10% level of significance to reject the claim that the mean monthly cost of joining a health club is \$25.

**39.** (a) The claim is "the mean expenditure per student in public elementary and secondary schools is more than \$12,000."

$H_0: \mu \le \$12{,}000$; $H_a: \mu > \$12{,}000$ (claim)

(b) $\bar{x} \approx 12,600.81$ $\qquad$ $s \approx 490.8821$

$$t = \frac{\bar{x} - \mu}{\frac{s}{\sqrt{n}}} = \frac{12,600.81 - 12,000}{\frac{490.8821}{\sqrt{16}}} \approx 4.896$$

$P$-value = {Area to right of $t = 4.896$} $\approx 0.000097$

(c) Because $P$-value $< 0.01 = \alpha$, reject $H_0$.

(d) There is enough evidence at the 1% level of significance to support the education publication's claim that the mean expenditure per student in public elementary and secondary schools is more than \$12,000.

**41.** $np = (40)(0.15) = 6 > 5$

$nq = (40)(0.85) = 34 > 5 \rightarrow$ can use normal distribution

$H_0: p = 0.15$ (claim); $H_a: p \neq 0.15$

$-z_0 = -1.96$, $z_0 = 1.96$; Rejection regions: $z < -1.96$, $z > 1.96$

$$z = \frac{\hat{p} - p}{\sqrt{\frac{pq}{n}}} = \frac{0.09 - 0.15}{\sqrt{\frac{(0.15)(0.85)}{40}}} \approx -1.063$$

Because $-1.96 < z < 1.96$, fail to reject $H_0$. There is not enough evidence at the 5% level of significance to reject the claim.

**43.** $np = (116)(0.65) = 75.4 > 5$

$nq = (116)(0.35) = 40.6 > 5 \rightarrow$ can use normal distribution

$H_0: p = 0.65$ (claim); $H_a: p \neq 0.65$

$-z_0 = -2.17$, $z_0 = 2.17$; Rejection regions: $z < -2.17$, $z > 2.17$

$$z = \frac{\hat{p} - p}{\sqrt{\frac{pq}{n}}} = \frac{0.76 - 0.65}{\sqrt{\frac{(0.65)(0.35)}{116}}} \approx 2.48$$

Because $z > 2.17$, reject $H_0$. There is not enough evidence at the 3% level of significance to support the claim.

**45.** (a) The claim is "over 60% of U.S. adults think that the federal government's bank bailouts were bad for the United States."

$H_0: p \leq 0.60$; $H_a: p > 0.60$ (claim)

(b) $z_0 = 2.33$; Rejection region: $z > 2.33$

(c) $\hat{p} = \frac{x}{n} = \frac{167}{298} \approx 0.560$

$$z = \frac{\hat{p} - p}{\sqrt{\frac{pq}{n}}} \approx \frac{0.560 - 0.60}{\sqrt{\frac{(0.60)(0.40)}{298}}} \approx -1.41$$

(d) Because $z < 2.33$, fail to reject $H_0$.

(e) There is not enough evidence at the 1% level of significance to support the claim that over 60% of U.S. adults think that the federal government's bank bailouts were bad for the United States.

**47.** Critical value: $\chi_0^2 = 30.144$; Rejection region: $\chi^2 > 30.144$

**49.** Critical value: $\chi_0^2 = 63.167$; Rejection region: $\chi^2 > 63.167$

**51.** $H_0: \sigma^2 \le 2$; $H_a: \sigma^2 > 2$ (claim)

$\chi_0^2 = 24.769$; Rejection region: $\chi^2 > 24.769$

$$\chi^2 = \frac{(n-1)s^2}{\sigma^2} = \frac{(17)(2.95)}{(2)} = 25.075$$

Because $\chi^2 > 24.769$, reject $H_0$. There is enough evidence at the 10% level of significance to support the claim.

**53.** $H_0: \sigma = 1.25$ (claim); $H_a: \sigma \ne 1.25$

$\chi_L^2 = 0.831$, $\chi_R^2 = 12.833$; Rejection regions: $\chi^2 < 0.831$, $\chi^2 > 12.833$

$$\chi^2 = \frac{(n-1)s^2}{\sigma^2} = \frac{(5)(1.03)^2}{(1.25)^2} \approx 3.395$$

Because $0.831 < \chi^2 < 12.833$, fail to reject $H_0$. There is not enough evidence at the 5% level of significance to reject the claim.

**55.** (a) The claim is "the variance of the bolt widths is at most 0.01."

$H_0: \sigma^2 \le 0.01$ (claim); $H_a: \sigma^2 > 0.01$

(b) $\chi_0^2 = 49.645$; Rejection region: $\chi^2 > 49.645$

(c) $\chi^2 = \dfrac{(n-1)s^2}{\sigma^2} = \dfrac{(27)(0.064)}{(0.01)} = 172.8$

(d) Because $\chi^2 > 49.645$, reject $H_0$.

(e) There is enough evidence at the 0.5% level of significance to reject the claim that the variance is at most 0.01.

**57.** $\chi_L^2 = 13.844$, $\chi_R^2 = 41.923$; Rejection regions: $\chi^2 < 13.844$, $\chi^2 > 41.923$

From Exercise 56, $\chi^2 = 43.94$.

You can reject $H_0$ at the 5% level of significance because $\chi^2 = 43.94 > 41.923$.

## CHAPTER 7 QUIZ SOLUTIONS

**1.** (a) The claim is "the mean hat size for a male is at least 7.25."

$H_0: \mu \ge 7.25$ (claim); $H_a: \mu < 7.25$

(b) Left-tailed because the alternative hypothesis contains $<$; $z$-test because $\sigma$ is known and the population is normally distributed.

(c) *Sample answer*: $z_0 = -2.33$; Rejection region: $z < -2.33$

$$z = \frac{\bar{x} - \mu}{\frac{\sigma}{\sqrt{n}}} = \frac{7.15 - 7.25}{\frac{0.27}{\sqrt{12}}} \approx -1.283$$

(d) Because $z > -2.33$, fail to reject $H_0$.

(e) There is not enough evidence at the 1% level of significance to reject the company's claim that the mean hat size for a male is at least 7.25.

2. (a) The claim is "the mean daily cost of meals and lodging for 2 adults traveling in Nevada is more than \$300."

$H_0: \mu \leq \$300$; $H_a: \mu > 300$ (claim)

(b) Right-tailed because the alternative hypothesis contains >; z-test because $\sigma$ is known and $n \geq 30$.

(c) *Sample answer:* $z_0 = 1.28$; Rejection region: $z > 1.28$

$$z = \frac{\bar{x} - \mu}{\frac{\sigma}{\sqrt{n}}} = \frac{316 - 300}{\frac{30}{\sqrt{35}}} \approx 3.156$$

(d) Because $z > 1.28$, reject $H_0$.

(e) There is enough evidence at the 10% level of significance to support the tourist agency's claim that the mean daily cost of meals and lodging for 2 adults traveling in Nevada is more than \$300.

3. (a) The claim is "the mean amount of earnings for full-time workers ages 25 to 34 with a master's degree is less than \$70,000."

$H_0: \mu \geq \$70,000$; $H_a: \mu < \$70,000$ (claim)

(b) Left-tailed because the alternative hypothesis contains <; t-test because $\sigma$ is unknown and the population is normally distributed.

(c) *Sample answer:* $t_0 = -1.761$; Rejection region: $t < -1.761$

$$t = \frac{\bar{x} - \mu}{\frac{s}{\sqrt{n}}} = \frac{66,231 - 70,000}{\frac{5945}{\sqrt{15}}} \approx -2.455$$

(d) Because $t < -1.761$, reject $H_0$.

(e) There is enough evidence at the 5% level of significance to support the agency's claim that the mean amount of earnings for full-time workers ages 25 to 34 with a master's degree is less than \$70,000.

4. (a) The claim is "program participants have a mean weight loss of at least 10 pounds after 1 month."

$H_0: \mu \geq 10$ (claim); $H_a: \mu < 10$

(b) Left-tailed because the alternative hypothesis contains <; t-test because $\sigma$ is unknown and $n \geq 30$.

(c) *Sample answer:* $t_0 = -2.462$; Rejection region: $t < -2.462$

$$t = \frac{\bar{x} - \mu}{\frac{s}{\sqrt{n}}} = \frac{8.78 - 10}{\frac{2.3615}{\sqrt{30}}} \approx -2.83$$

(d) Because $t < -2.462$, reject $H_0$.

(e) There is enough evidence at the 1% level of significance to reject the claim that the program participants have a mean weight loss of at least 10 pounds after 1 month.

5. (a) The claim is "less than 10% of microwaves need repair during the first 5 years of use."
$H_0: p \geq 0.10$; $H_a: p < 0.10$ (claim)

(b) Left-tailed because the alternative hypothesis contains $<$; $z$-test because $np \geq 5$ and $nq \geq 5$.

(c) *Sample answer:* $z_0 = -1.645$; Rejection region: $z < -1.645$

$$z = \frac{\hat{p} - p}{\sqrt{\dfrac{pq}{n}}} = \frac{0.13 - 0.10}{\sqrt{\dfrac{(0.10)(0.90)}{57}}} \approx 0.75$$

(d) Because $z > -1.645$, fail to reject $H_0$.

(e) There is not enough evidence at the 5% level of significance to support the claim that less than 10% of microwaves need repair during the first 5 years of use.

6. (a) The claim is "the standard deviation of SAT critical reading test scores is 114."
$H_0: \sigma = 114$ (claim); $H_a: \sigma \neq 114$

(b) Two-tailed because the alternative hypothesis contains $\neq$; chi-square test because the test is for a standard deviation and the population is normally distributed.

(c) *Sample answer:* $\chi_L^2 = 9.390$, $\chi_R^2 = 28.869$; Rejection regions: $\chi^2 < 9.390$, $\chi^2 > 28.869$

$$\chi^2 = \frac{(n-1)s^2}{\sigma^2} = \frac{(18)(143)^2}{(114)^2} \approx 28.323$$

(d) Because $9.390 < \chi^2 < 28.869$, fail to reject $H_0$.

(e) There is not enough evidence at the 10% level of significance to reject the school administrator's claim that the standard deviation of SAT critical reading test scores is 114.

# Hypothesis Testing with Two Samples

## 8.1 Try It Yourself Solutions

**Note:** Answers may differ due to rounding.

**1a.** (1) Independent
   (2) Dependent
**b.** (1) Because each sample represents blood pressures of different individuals, and it is not possible to form a pairing between the members of the samples.
   (2) Because the samples represent exam scores of the same students, the samples can be paired with respect to each student.

**2.** The claim is "there is a difference in the mean annual wages for forensic science technicians working for local and state governments."
**a.** $H_0 : \mu_1 = \mu_2$; $H_a : \mu_1 \neq \mu_2$ (claim)
**b,** $\alpha = 0.10$
**c.** $-z_0 = -1.645$, $z_0 = 1.645$; Rejection regions: $z < -1.645$, $z > 1.645$
**d.** $z = \dfrac{(\bar{x}_1 - \bar{x}_2) - (\mu_1 - \mu_2)}{\sqrt{\dfrac{\sigma_1^2}{n_1} + \dfrac{\sigma_2^2}{n_2}}} = \dfrac{(55,950 - 51,100) - (0)}{\sqrt{\dfrac{(6200)^2}{100} + \dfrac{(5575)^2}{100}}} \approx 5.817$

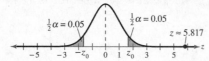

**e.** Reject $H_0$.

**f.** There is enough evidence at the 10% level of significance to support the claim that there is a difference in the mean annual wages for forensic science technicians working for local and state governments.

**3a.** $z = \dfrac{(\bar{x}_1 - \bar{x}_2) - (\mu_1 - \mu_2)}{\sqrt{\dfrac{\sigma_1^2}{n_1} + \dfrac{\sigma_2^2}{n_2}}} = \dfrac{(296 - 293) - (0)}{\sqrt{\dfrac{(24)^2}{15} + \dfrac{(19)^2}{20}}} \approx 0.399$

   $\rightarrow$ $P$-value = {area right of $z = 0.399$} $\approx 0.3448$
**b.** Because $P$-value $> 0.05 = \alpha$, fail to reject $H_0$.
**c.** There is not enough evidence at the 5% level of significance to support the travel agency's claim that the average daily cost of meals and lodging for vacationing in Alaska is greater than the average daily cost for vacationing in Colorado.

## 8.1 EXERCISE SOLUTIONS

1.  Two samples are dependent if each member of one sample corresponds to a member of the other sample. Example: The weights of 22 people before starting an exercise program and the weights of the same 22 people 6 weeks after starting the exercise program.

    Two samples are independent if the sample selected from one population is not related to the sample from the other population. Example: The weights of 25 cats and the weights of 20 dogs.

3.  Use $P$-values.

5.  Dependent because the same football players were sampled.

7.  Independent because different boats were sampled.

9.  Because $z \approx 2.96 > 1.96$ and $P \approx 0.0031 < 0.05,$ reject $H_0$.

11. $H_0 : \mu_1 = \mu_2$ (claim); $H_a : \mu_1 \neq \mu_2$
    Rejection region: $z < -2.575,\ z > 2.575$
    $$z = \frac{(\bar{x}_1 - \bar{x}_2) - (\mu_1 - \mu_2)}{\sqrt{\dfrac{\sigma_1^2}{n_1} + \dfrac{\sigma_2^2}{n_2}}} = \frac{(16 - 14) - (0)}{\sqrt{\dfrac{(3.4)^2}{29} + \dfrac{(1.5)^2}{28}}} \approx 2.89$$
    Because $z > 2.575$, reject $H_0$. There is enough evidence at the 1% level of significance to reject the claim.

13. $H_0 : \mu_1 \geq \mu_2$; $H_a : \mu_1 < \mu_2$ (claim)
    Rejection region: $z < -1.645$
    $$z = \frac{(\bar{x}_1 - \bar{x}_2) - (\mu_1 - \mu_2)}{\sqrt{\dfrac{\sigma_1^2}{n_1} + \dfrac{\sigma_2^2}{n_2}}} = \frac{(2435 - 2432) - (0)}{\sqrt{\dfrac{(75)^2}{35} + \dfrac{(105)^2}{90}}} \approx 0.18$$
    Because $z > -1.645$, fail to reject $H_0$. There is not enough evidence at the 5% level of significance to support the claim.

15. (a) The claim is "the mean braking distances are different for the two types of tires."
    $H_0 : \mu_1 = \mu_2$; $H_a : \mu_1 \neq \mu_2$ (claim)
    (b) $-z_0 = -1.645,\ z_0 = 1.645$; Rejection regions: $z < -1.645,\ z > 1.645$
    (c) $z = \dfrac{(\bar{x}_1 - \bar{x}_2) - (\mu_1 - \mu_2)}{\sqrt{\dfrac{\sigma_1^2}{n_1} + \dfrac{\sigma_2^2}{n_2}}} = \dfrac{(42 - 45) - (0)}{\sqrt{\dfrac{(4.7)^2}{35} + \dfrac{(4.3)^2}{35}}} \approx -2.786$
    (d) Because $z < -1.645$, reject $H_0$.
    (e) There is enough evidence at the 10% level of significance to support the claim that the mean braking distances are different for the two types of tires.

17. (a) The claim is "the wind speed in Region A is less than the wind speed in Region B."
    $H_0 : \mu_1 \geq \mu_2$; $H_a : \mu_1 < \mu_2$ (claim)

(b) $z_0 > -1.645$; Rejection region: $z < -1.645$

(c) $z = \dfrac{(\bar{x}_1 - \bar{x}_2) - (\mu_1 - \mu_2)}{\sqrt{\dfrac{\sigma_1^2}{n_1} + \dfrac{\sigma_2^2}{n_2}}} = \dfrac{(14 - 15.1) - (0)}{\sqrt{\dfrac{(2.9)^2}{60} + \dfrac{(3.3)^2}{60}}} \approx -1.94$

(d) Because, $z < -1.645$ reject $H_0$.

(e) There is enough evidence at the 5% level of significance to conclude that the wind speed in Region A is less than the wind speed in Region B.

19. (a) The claim is "male and female high school students have equal ACT scores."

    $H_0 : \mu_1 = \mu_2$ (claim); $H_a : \mu_1 \neq \mu_2$

    (b) $-z_0 = -2.575$, $z_0 = 2.575$; Rejection regions: $z < -2.575$, $z > 2.575$

    (c) $z = \dfrac{(\bar{x}_1 - \bar{x}_2) - (\mu_1 - \mu_2)}{\sqrt{\dfrac{\sigma_1^2}{n_1} + \dfrac{\sigma_2^2}{n_2}}} = \dfrac{(21.0 - 20.8) - (0)}{\sqrt{\dfrac{(5.0)^2}{43} + \dfrac{(4.7)^2}{56}}} \approx 0.202$

    (d) Because $-2.575 < z < 2.575$, fail to reject $H_0$.

    (e) There is not enough evidence at the 1% level of significance to reject the claim that the male and female high school students have equal ACT scores.

21. The claim is "the mean home sales price in Spring, Texas is the same as in Austin, Texas."

    (a) $H_0 : \mu_1 = \mu_2$ (claim); $H_a : \mu_1 \neq \mu_2$

    (b) $-z_0 = -2.575$, $z_0 = 2.575$; Rejection regions: $z < -2.575$, $z > 2.575$

    (c) $z = \dfrac{(\bar{x}_1 - \bar{x}_2) - (\mu_1 - \mu_2)}{\sqrt{\dfrac{\sigma_1^2}{n_1} + \dfrac{\sigma_2^2}{n_2}}} = \dfrac{(127{,}414 - 112{,}301) - (0)}{\sqrt{\dfrac{(25{,}875)^2}{25} + \dfrac{(27{,}110)^2}{25}}} \approx 2.02$

    (d) Because $-2.575 < z < 2.575$, fail to reject $H_0$.

    (e) There is not enough evidence at the 1% level of significance to reject the claim that the mean home sales price in Spring, Texas is the same as in Austin, Texas.

23. (a) The claim is "children ages 6–17 spent more time watching television in 1981 than children ages 6–17 do today."

    $H_0 : \mu_1 \leq \mu_2$; $H_a : \mu_1 > \mu_2$ (claim)

    (b) $z_0 = 1.645$; Rejection region: $z > 1.645$

    (c) $\bar{x}_1 = 2.13$, $n_1 = 30$

    $\bar{x}_2 \approx 1.76$, $n_2 = 30$

    $z = \dfrac{(\bar{x}_1 - \bar{x}_2) - (\mu_1 - \mu_2)}{\sqrt{\dfrac{\sigma_1^2}{n_1} + \dfrac{\sigma_2^2}{n_2}}} = \dfrac{(2.13 - 1.76) - (0)}{\sqrt{\dfrac{(0.6)^2}{30} + \dfrac{(0.5)^2}{30}}} \approx 2.59$

    (d) Because $z > 1.645$, reject $H_0$.

    (e) There is enough evidence at the 5% level of significance to support the claim that children ages 6–17 spent more time watching television in 1981 than children ages 6–17 do today.

25. They are equivalent through algebraic manipulation of the equation.

    $\mu_1 = \mu_2 \rightarrow \mu_1 - \mu_2 = 0$

**27.** $H_0 : \mu_1 - \mu_2 \leq 10{,}000; H_a : \mu_1 - \mu_2 > 10{,}000$ (claim)

$z_0 = 1.645$; Rejection region: $z > 1.645$

$$z = \frac{(\overline{x}_1 - \overline{x}_2) - (\mu_1 - \mu_2)}{\sqrt{\dfrac{\sigma_1^2}{n_1} + \dfrac{\sigma_2^2}{n_2}}} = \frac{(102{,}650 - 85{,}430) - (10{,}000)}{\sqrt{\dfrac{(8795)^2}{42} + \dfrac{(9250)^2}{38}}} \approx 3.569$$

Because $z > 1.645$, reject $H_0$. There is enough evidence at the 5% level of significance to support the claim that the difference in mean annual salaries of microbiologists in Maryland and California is more than $10,000.

**29.**

$$(\overline{x}_1 - \overline{x}_2) - z_c \sqrt{\frac{\sigma_1^2}{n_1} + \frac{\sigma_2^2}{n_2}} < \mu_1 - \mu_2 < (\overline{x}_1 - \overline{x}_2) + z_c \sqrt{\frac{\sigma_1^2}{n_1} + \frac{\sigma_2^2}{n_2}}$$

$$(102{,}650 - 85{,}430) - 1.96 \sqrt{\frac{(8795)^2}{42} + \frac{(9250)^2}{38}} < \mu_1 - \mu_2 < (102{,}650 - 85{,}430) + 1.96 \sqrt{\frac{(8795)^2}{42} + \frac{(9250)^2}{38}}$$

$$17{,}220 - 1.96\sqrt{4{,}093{,}360} < \mu_1 - \mu_2 < 17{,}220 + 1.96\sqrt{4{,}093{,}360}$$

$$\$13{,}255 < \mu_1 - \mu_2 < \$21{,}185$$

## 8.2 TESTING THE DIFFERENCE BETWEEN MEANS (SMALL INDEPENDENT SAMPLES)

### 8.2 Try It Yourself Solutions

**1a.** The claim is "there is a difference in the mean annual earnings based on level of education."

$H_0 : \mu_1 = \mu_2; H_a : \mu_1 \neq \mu_2$ (claim)

**b.** $\alpha = 0.01$; d.f. $= \min\{n_1 - 1, n_2 - 1\} = \min\{19 - 1, 16 - 1\} = 15$

**c.** $-t_0 = -2.947$, $t_0 = 2.947$; Rejection regions: $t < -2.947$, $t > 2.947$

**d.** $t = \dfrac{(\overline{x}_1 - \overline{x}_2) - (\mu_1 - \mu_2)}{\sqrt{\dfrac{s_1^2}{n_1} + \dfrac{s_2^2}{n_2}}} = \dfrac{(32{,}493 - 40{,}907) - (0)}{\sqrt{\dfrac{(3118)^2}{19} + \dfrac{(6162)^2}{16}}} \approx -4.95$

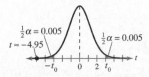

**e.** Because $t < -2.947$, reject $H_0$.

**f.** There is enough evidence that "there is a difference in the mean annual earnings based on level of education."

**2a.** The claim is "the mean operating cost per mile of the manufacturer's minivans is less than that of its leading competitor."

$H_0 : \mu_1 \geq \mu_2; H_a : \mu_1 < \mu_2$ (claim)

**b.** $\alpha = 0.10$; d.f. $= n_1 + n_2 - 2 = 34 + 38 - 2 = 70$

**c.** $t_0 = -1.294$; Rejection regions: $t < -1.294$

**d.** $t = \dfrac{(\bar{x}_1 - \bar{x}_2) - (\mu_1 - \mu_2)}{\sqrt{\dfrac{(n-1)s_1^2 + (n_2-1)s_2^2}{n_1 + n_2 - 2}} \sqrt{\dfrac{1}{n_1} + \dfrac{1}{n_2}}} = \dfrac{(0.56 - 0.58) - (0)}{\sqrt{\dfrac{(34-1)(0.08)^2 + (38-1)(0.07)^2}{34 + 38 - 2}} \sqrt{\dfrac{1}{34} + \dfrac{1}{38}}} \approx -1.13$

$\alpha = 0.10$

$t \approx -1.13$

**e.** Because $t > -1.294$, fail to reject $H_0$.

**f.** There is not enough evidence at the 10% level of significance to support the manufacturer's claim that the mean operating cost per mile of its minivans is less than that of its leading competitor.

## 8.2 EXERCISE SOLUTIONS

**1.** (1) The population standard deviations are unknown.
(2) The samples are randomly selected.
(3) The samples are independent
(4) The populations are normally distributed or each sample size is at least 30.

**3.** (a) d.f. $= n_1 + n_2 - 2 = 23$
$-t_0 = -1.714$, $t_0 = 1.714$
(b) d.f. $= \min\{n_1 - 1,\ n_2 - 1\} = 10$
$-t_0 = -1.812$, $t_0 = 1.812$

**5.** (a) d.f. $= n_1 + n_2 - 2 = 16$
$t_0 = -1.746$
(b) d.f. $= \min\{n_1 - 1,\ n_2 - 1\} = 6$
$t_0 = -1.943$

**7.** (a) d.f. $= n_1 + n_2 - 2 = 19$
$t_0 = 1.729$
(b) d.f. $= \min\{n_1 - 1,\ n_2 - 1\} = 7$
$t_0 = 1.895$

**9.** $H_0 : \mu_1 = \mu_2$ (claim); $H_a : \mu_1 \neq \mu_2$
d.f. $= n_1 + n_2 - 2 = 27$
Rejection regions: $t < -2.771$, $t > 2.771$

$t = \dfrac{(\bar{x}_1 - \bar{x}_2) - (\mu_1 - \mu_2)}{\sqrt{\dfrac{(n-1)s_1^2 + (n_2-1)s_2^2}{n_1 + n_2 - 2}} \sqrt{\dfrac{1}{n_1} + \dfrac{1}{n_2}}} = \dfrac{(33.7 - 35.5) - (0)}{\sqrt{\dfrac{(12-1)(3.5)^2 + (17-1)(2.2)^2}{12 + 17 - 2}} \sqrt{\dfrac{1}{12} + \dfrac{1}{17}}} \approx -1.70$

Because $-2.771 < t < 2.771$, fail to reject $H_0$. There is not enough evidence at the 1% level of significance to reject the claim.

**11.** $H_0 : \mu_1 \leq \mu_2$ (claim); $H_a : \mu_1 > \mu_2$
d.f. $= \min\{n_1 - 1,\ n_2 - 1\} = 9$
Rejection region: $t > 1.833$

$$t = \frac{(\bar{x}_1 - \bar{x}_2) - (\mu_1 - \mu_2)}{\sqrt{\dfrac{s_1^2}{n_1} + \dfrac{s_2^2}{n_2}}} = \frac{(2410 - 2305) - (0)}{\sqrt{\dfrac{(175)^2}{13} + \dfrac{(52)^2}{10}}} \approx 2.05$$

Because $t > 1.833$, reject $H_0$. There is enough evidence at the 5% level of significance to reject the claim.

13. (a) The claim is "the mean annual costs of routine veterinarian visits for dogs and cats are the same."

$H_0 : \mu_1 = \mu_2$ (claim); $H_a : \mu_1 \neq \mu_2$

(b) d.f. $= \min\{n_1 - 1, \ n_2 - 1\} = 15$

$-t_0 = -1.753$, $t_0 = 1.753$; Rejection regions: $t < -1.753$, $t > 1.753$

(c) $t = \dfrac{(\bar{x}_1 - \bar{x}_2) - (\mu_1 - \mu_2)}{\sqrt{\dfrac{s_1^2}{n_1} + \dfrac{s_2^2}{n_2}}} = \dfrac{(239 - 203) - (0)}{\sqrt{\dfrac{(32)^2}{16} + \dfrac{(21)^2}{18}}} \approx 3.83$

(d) Because $t > 1.753$, reject $H_0$.

(e) There is enough evidence at the 10% level of significance to reject the claim that the mean annual costs of routine veterinarians visits for dogs and cats are the same.

15. (a) The claim is "the mean length of mature female pink seaperch is different in fall and winter."

$H_0 : \mu_1 = \mu_2$; $H_a : \mu_1 \neq \mu_2$ (claim)

(b) d.f. $= n_1 + n_2 - 2 = 55$

$-t_0 \approx -2.678$, $t_0 \approx 2.678$; Rejection regions: $t < -2.678$, $t > 2.678$

(c) $t = \dfrac{(\bar{x}_1 - \bar{x}_2) - (\mu_1 - \mu_2)}{\sqrt{\dfrac{(n-1)s_1^2 + (n_2-1)s_2^2}{n_1 + n_2 - 2}} \sqrt{\dfrac{1}{n_1} + \dfrac{1}{n_2}}} = \dfrac{(127 - 117) - (0)}{\sqrt{\dfrac{(26-1)(14)^2 + (31-1)(9)^2}{26 + 31 - 2}} \sqrt{\dfrac{1}{26} + \dfrac{1}{31}}} \approx 3.26$

(d) Because $t > 2.678$, reject $H_0$.

(e) There is enough evidence at the 1% level of significance to support the claim that the mean length of mature female pink seaperch is different in fall and winter.

17. (a) The claim is "the mean household income is greater in Allegheny County than it is in Erie County."

$H_0 : \mu_1 \leq \mu_2$; $H_a : \mu_1 > \mu_2$ (claim)

(b) d.f. $= \min\{n_1 - 1, \ n_2 - 1\} = 14$

$t_0 = 1.761$; Rejection region: $t > 1.761$

(c) $t = \dfrac{(\bar{x}_1 - \bar{x}_2) - (0)}{\sqrt{\dfrac{s_1^2}{n_1} + \dfrac{s_2^2}{n_2}}} = \dfrac{(49,700 - 42,000) - (0)}{\sqrt{\dfrac{(8800)^2}{19} + \dfrac{(5100)^2}{15}}} \approx 3.19$

(d) Because $t > 1.761$, reject $H_0$.

(e) There is enough evidence at the 5% level of significance to support the claim that the mean household income is greater in Allegheny County than it is in Erie County.

19. (a) The claim is "the new treatment makes a difference in the tensile strength of steel bars."

$H_0 : \mu_1 = \mu_2$; $H_a : \mu_1 \neq \mu_2$ (claim)

(b) d.f. $= n_1 + n_2 - 2 = 21$

$-t_0 = -2.831$, $t_0 = 2.831$; Rejection regions: $t < -2.831$, $t > 2.831$

(c) $\bar{x}_1 = 368.3$, $s_1 \approx 22.301$, $n_1 = 10$

$\bar{x}_2 \approx 389.538$, $s_2 \approx 14.512$, $n_2 = 13$

$$t = \frac{(\bar{x}_1 - \bar{x}_2) - (\mu_1 - \mu_2)}{\sqrt{\dfrac{(n-1)s_1^2 + (n_2-1)s_2^2}{n_1 + n_2 - 2}}\sqrt{\dfrac{1}{n_1} + \dfrac{1}{n_2}}} \approx \frac{(368.3 - 389.538) - (0)}{\sqrt{\dfrac{(10-1)(22.301)^2 + (13-1)(14.512)^2}{10 + 13 - 2}}\sqrt{\dfrac{1}{10} + \dfrac{1}{13}}} \approx -2.76$$

(d) Because $-2.831 < t < 2.831$, fail to reject $H_0$.

(e) There is not enough evidence at the 1% level of significance to support the claim that the new treatment makes a difference in the tensile strength of steel bars.

21. (a) The claim is "the new method of teaching reading produces higher reading test scores than the old method"

$H_0 : \mu_1 \geq \mu_2$; $H_a : \mu_1 < \mu_2$ (claim)

(b) d.f. $= n_1 + n_2 - 2 = 42$

$t_0 \approx -1.303$; Rejection region: $t < -1.303$

(c) $\bar{x}_1 \approx 56.684$, $s_1 \approx 6.961$, $n_1 = 19$

$\bar{x}_2 \approx 67.4$, $s_2 \approx 9.014$, $n_2 = 25$

$$t = \frac{(\bar{x}_1 - \bar{x}_2) - (\mu_1 - \mu_2)}{\sqrt{\dfrac{(n-1)s_1^2 + (n_2-1)s_2^2}{n_1 + n_2 - 2}}\sqrt{\dfrac{1}{n_1} + \dfrac{1}{n_2}}} \approx \frac{(56.684 - 67.4) - (0)}{\sqrt{\dfrac{(19-1)(6.961)^2 + (25-1)(9.014)^2}{19 + 25 - 2}}\sqrt{\dfrac{1}{19} + \dfrac{1}{25}}} \approx -4.295$$

(d) Because $t < -1.303$, reject $H_0$.

(e) There is enough evidence at the 10% level of significance to support the claim that the new method of teaching reading produces higher reading test scores than the old method.

23. $(\bar{x}_1 - \bar{x}_2) \pm t_c \sqrt{\dfrac{s_1^2}{n_1} + \dfrac{s_2^2}{n_2}} \rightarrow (267 - 244) \pm 2.132 \sqrt{\dfrac{(6)^2}{9} + \dfrac{(12)^2}{5}}$

$\rightarrow 23 \pm 12.21 \rightarrow 10.78 < \mu_1 - \mu_2 < 35.21 \rightarrow 11 < \mu_1 - \mu_2 < 35$

25. $\hat{\sigma} = \sqrt{\dfrac{(n-1)s_1^2 + (n_2-1)s_2^2}{n_1 + n_2 - 2}} = \sqrt{\dfrac{(21-1)(166)^2 + (11-1)(204)^2}{21 + 11 - 2}} \approx 179.56$

$(\bar{x}_1 - \bar{x}_2) \pm t_c \hat{\sigma} \sqrt{\dfrac{1}{n_1} + \dfrac{1}{n_2}} \rightarrow (1805 - 1629) \pm 2.042 \cdot 179.56 \sqrt{\dfrac{1}{21} + \dfrac{1}{11}}$

$\rightarrow 176 \pm 136.47 \rightarrow 39.53 < \mu_1 - \mu_2 < 312.47 \rightarrow 40 < \mu_1 - \mu_2 < 312$

## 8.3 TESTING THE DIFFERENCE BETWEEN MEANS (DEPENDENT SAMPLES)

### 8.3 Try It Yourself Solutions

**1a.** The claim is "athletes can decrease their times in the 40-yard dash."

$H_0 : \mu_d \leq 0$; $H_a : \mu_d > 0$ (claim)

**b.** $\alpha = 0.05$, d.f. $= n - 1 = 11$

**c.** $t_0 = 1.796$; Rejection region: $t > 1.796$

**d.**

| Before | After | $d$ | $d^2$ |
|--------|-------|------|--------|
| 4.85 | 4.78 | 0.07 | 0.0049 |
| 4.90 | 4.90 | 0.00 | 0.0000 |
| 5.08 | 5.05 | 0.03 | 0.0009 |
| 4.72 | 4.65 | 0.07 | 0.0049 |
| 4.62 | 4.64 | −0.02 | 0.0004 |
| 4.54 | 4.50 | 0.04 | 0.0016 |
| 5.25 | 5.24 | 0.01 | 0.0001 |
| 5.18 | 5.27 | −0.09 | 0.0081 |
| 4.81 | 4.75 | 0.06 | 0.0036 |
| 4.57 | 4.43 | 0.14 | 0.0196 |
| 4.63 | 4.61 | 0.02 | 0.0004 |
| 4.77 | 4.82 | −0.05 | 0.0025 |
| | | $\sum d = 0.28$ | $\sum d^2 = 0.047$ |

$$\bar{d} = \frac{\sum d}{n} = \frac{0.28}{12} \approx 0.0233$$

$$s_d = \sqrt{\frac{\left(\sum d^2\right) - \left[\frac{\left(\sum d\right)^2}{n}\right]}{n-1}} = \sqrt{\frac{0.047 - \frac{(0.28)^2}{12}}{11}} \approx 0.0607$$

**e.** $t = \dfrac{\bar{d} - \mu_d}{\dfrac{s_d}{\sqrt{n}}} \approx \dfrac{0.0233 - 0}{\dfrac{0.0607}{\sqrt{12}}} \approx 1.330$

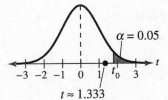

$\alpha = 0.05$

$t \approx 1.333$

**f.** Because $t < 1.796$, fail to reject $H_0$.

**g.** There is not enough evidence at the 5% level of significance to support the claim that athletes can decrease their times in the 40-yard dash using new strength shoes.

**2a.** The claim is "the drug changes the body's temperature."

$H_0 : \mu_d = 0$; $H_a : \mu_d \neq 0$ (claim)

**b.** $\alpha = 0.05$, d.f. $= n - 1 = 6$

**c.** $-t_0 = -2.447$, $t_0 = 2.447$; Rejection regions: $t < -2.447$, $t > 2.447$

**d.**

| Before | After | $d$ | $d^2$ |
|--------|-------|-----|-------|
| 101.8 | 99.2 | 2.6 | 6.76 |
| 98.5 | 98.4 | 0.1 | 0.01 |
| 98.1 | 98.2 | −0.1 | 0.01 |
| 99.4 | 99 | 0.4 | 0.16 |
| 98.9 | 98.6 | 0.3 | 0.09 |
| 100.2 | 99.7 | 0.5 | 0.25 |
| 97.9 | 97.8 | 0.1 | 0.01 |
| | | $\sum d = 3.9$ | $\sum d^2 = 7.29$ |

$$\overline{d} = \frac{\sum d}{n} = \frac{3.9}{7} \approx 0.5771$$

$$s_d = \sqrt{\frac{\sum d^2 - \left[\frac{\left(\sum d\right)^2}{n}\right]}{n-1}} = \sqrt{\frac{7.29 - \frac{(3.9)^2}{7}}{6}} \approx 0.9235$$

**e.** $t = \dfrac{\overline{d} - \mu_d}{\frac{s_d}{\sqrt{n}}} \approx \dfrac{0.5571 - 0}{\frac{0.9235}{\sqrt{7}}} \approx 1.596$

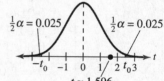

$\frac{1}{2}\alpha = 0.025$   $\frac{1}{2}\alpha = 0.025$

$t \approx 1.596$

**f.** Because $-2.447 < t < 2.447$, fail to reject $H_0$.

**g.** There is not enough evidence at the 5% level of significance to conclude that the drug changes the body's temperature.

## 8.3 EXERCISE SOLUTIONS

**1.** (1) The samples are randomly selected.
   (2) The samples are dependent.
   (3) The populations are normally distributed or the number $n$ of pairs of data is at least 30.

**3.** $H_0 : \mu_d \geq 0$; $H_a : \mu_d < 0$ (claim)

   $\alpha = 0.05$ and d.f. $= n - 1 = 13$

   $t_0 = -1.771$; Rejection region: $t < -1.771$

   $t = \dfrac{\overline{d} - \mu_d}{\frac{s_d}{\sqrt{n}}} = \dfrac{1.5 - 0}{\frac{3.2}{\sqrt{14}}} \approx 1.754$

   Because $t > -1.771$, fail to reject $H_0$. There is not enough evidence at the 5% level of significance to support the claim.

5.  $H_0 : \mu_d \leq 0$ (claim); $H_a : \mu_d > 0$

    $\alpha = 0.10$ and d.f. $= n - 1 = 15$

    $t_0 = 1.341$; Rejection region: $t > 1.341$

    $t = \dfrac{\overline{d} - \mu_d}{\dfrac{s_d}{\sqrt{n}}} = \dfrac{6.5 - 0}{\dfrac{9.54}{\sqrt{16}}} \approx 2.725$

    Because $t > 1.341$, reject $H_0$. There is enough evidence at the 10% level of significance to reject the claim.

7.  $H_0 : \mu_d \geq 0$ (claim); $H_a : \mu_d < 0$

    $\alpha = 0.01$ and d.f. $= n - 1 = 14$

    $t_0 = -2.624$; Rejection region: $t < -2.624$

    $t = \dfrac{\overline{d} - \mu_d}{\dfrac{s_d}{\sqrt{n}}} = \dfrac{-2.3 - 0}{\dfrac{1.2}{\sqrt{15}}} \approx -7.423$

    Because $t < -2.624$, reject $H_0$. There is enough evidence at the 1% level of significance to reject the claim.

9.  (a)  The claim is "pneumonia causes weight loss in mice."
        $H_0 : \mu_d \leq 0; H_a : \mu_d > 0$ (claim)

    (b)  $t_0 = 3.365$; Rejection region: $t > 3.365$

    (c)  $\overline{d} = 1.05$ and $s_d \approx 0.345$

    (d)  $t = \dfrac{\overline{d} - \mu_d}{\dfrac{s_d}{\sqrt{n}}} = \dfrac{1.05 - 0}{\dfrac{0.345}{\sqrt{6}}} \approx 7.455$

    (e)  Because $t > 3.365$, reject $H_0$.

    (f)  There is enough evidence at the 1% level of significance to support the claim that pneumonia causes weight loss in mice."

11. (a)  The claim is "a post-lunch nap decreases the amount of time it takes males to sprint 20 meters after a night with only 4 hours of sleep."
        $H_0 : \mu_d \leq 0; H_a : \mu_d > 0$ (claim)

    (b)  $t_0 = 2.821$; Rejection region: $t > 2.821$

    (c)  $\overline{d} = 0.097$ and $s_d \approx 0.043$

    (d)  $t = \dfrac{\overline{d} - \mu_d}{\dfrac{s_d}{\sqrt{n}}} = \dfrac{0.097 - 0}{\dfrac{0.043}{\sqrt{10}}} \approx 7.134$

    (e)  Because $t > 2.821$, reject $H_0$.

    (f)  There is enough evidence at the 1% level of significance to support the claim that a post-lunch nap decreases the amount of time it takes males to sprint 20 meters after a night with only 4 hours of sleep.

**13.** (a) The claim is "soft tissue therapy and spinal manipulation help to reduce the lengths of time patients suffer from headaches."

$H_0 : \mu_d \leq 0; H_a : \mu_d > 0$ (claim)

(b) $t_0 = 2.764$; Rejection region; $t > 2.764$

(c) $\bar{d} = 1.255$ and $s_d \approx 0.441$

(d) $t = \dfrac{\bar{d} - \mu_d}{\frac{s_d}{\sqrt{n}}} = \dfrac{1.255 - 0}{\frac{0.441}{\sqrt{11}}} \approx 9.438$

(e) Because $t > 2.764$, reject $H_0$.

(f) There is enough evidence at the 1% level of significance to support the claim that soft tissue therapy and spinal manipulation help reduce the length of time patients suffer from headaches.

**15.** (a) The claim is "high intensity power training decreases the body fat percentages of females."

$H_0 : \mu_d \leq 0; H_a : \mu_d > 0$ (claim)

(b) $t_0 = 1.895$; Rejection region: $t > 1.895$

(c) $\bar{d} = 2.475$ and $s_d \approx 2.172$

(d) $t = \dfrac{\bar{d} - \mu_d}{\frac{s_d}{\sqrt{n}}} = \dfrac{2.475 - 0}{\frac{2.172}{\sqrt{8}}} \approx 3.223$

(e) Because $t > 1.895$, reject $H_0$.

(f) There is enough evidence at the 5% level of significance to support the claim that high intensity power training decreases the body fat percentages of females.

**17.** (a) The claim is "the product ratings have changed from last year to this year."

$H_0 : \mu_d = 0; H_a : \mu_d \neq 0$ (claim)

(b) $-t_0 = -2.365, t_0 = 2.365$; Rejection regions: $t < -2.365, t > 2.365$

(c) $\bar{d} = -1$ and $s_d \approx 1.309$

(d) $t = \dfrac{\bar{d} - \mu_d}{\frac{s_d}{\sqrt{n}}} = \dfrac{-1 - 0}{\frac{1.309}{\sqrt{8}}} \approx -2.161$

(e) Because $-2.365 < t < 2.365$, fail to reject $H_0$.

(f) There is not enough evidence at the 5% level of significance to support the claim that the product ratings have changed from last year to this year.

**19.** (a) The claim is "eating new cereal as part of a daily diet lowers total blood cholesterol levels."

$H_0 : \mu_d \leq 0; H_a : \mu_d > 0$ (claim)

(b) $t_0 = 1.943$; Rejection region: $t > 1.943$

(c) $\bar{d} \approx 2.857$ and $s_d \approx 4.451$

(d) $t = \dfrac{\bar{d} - \mu_d}{\frac{s_d}{\sqrt{n}}} = \dfrac{2.857 - 0}{\frac{4.451}{\sqrt{7}}} \approx 1.698$

(e) Because $t < 1.943$, fail to reject $H_0$.

(f) There is not enough evidence at the 5% level of significance to support the claim that eating new cereal as part of a daily diet lowers total blood cholesterol levels.

**21.** Yes; $P \approx 0.0073 < 0.05$, so you reject $H_0$.

**23.** $\bar{d} = -1.525$ and $s_d \approx 0.542$

$$\bar{d} - t_{\alpha/2} \frac{s_d}{\sqrt{n}} < \mu_d < \bar{d} - t_{\alpha/2} \frac{s_d}{\sqrt{n}}$$

$$-1.525 - 1.753\left(\frac{0.542}{\sqrt{16}}\right) < \mu_d < -1.525 + 1.753\left(\frac{0.542}{\sqrt{16}}\right)$$

$$-1.525 - 0.238 < \mu_d < -1.525 + 0.238$$

$$-1.76 < \mu_d < -1.29$$

## 8.4 TESTING THE DIFFERENCE BETWEEN PROPORTIONS

## 8.4 Try It Yourself Solutions

**1a.** $\bar{p} = \dfrac{x_1 + x_2}{n_1 + n_2} = \dfrac{367 + 6290}{1593 + 29948} \approx 0.2111, \ \bar{q} = 1 - \bar{p} \approx 0.7889$

**b.** $n_1\bar{p} = 1593(0.2111) \approx 336.3 > 5, \ n_1\bar{q} = 1593(0.7889) \approx 1256.7 > 5$

$n_2\bar{p} = 29{,}948(0.2111) \approx 6322.0 > 5, \ n_2\bar{q} = 29{,}948(0.7889) \approx 23{,}626.0 > 5$

**c.** The claim is "there is a difference between the proportion 40- to 49-year olds who are yoga users and the proportion of 40- to 49-year-olds who are non-yoga users."

$H_0 : p_1 = p_2; \ H_a : p_1 \neq p_2$ (claim)

**d.** $\alpha = 0.05$

**e.** $-z_0 = -1.96, \ z_0 = 1.96;$ Rejection regions: $z < -1.96, \ z > 1.96$

**f.** $z = \dfrac{(\hat{p}_1 - \hat{p}_2) - (p_1 - p_2)}{\sqrt{\bar{p}\bar{q}\left(\dfrac{1}{n_1} + \dfrac{1}{n_2}\right)}} = \dfrac{(0.230 - 0.210) - (0)}{\sqrt{0.211(0.789)\left(\dfrac{1}{1593} + \dfrac{1}{29{,}948}\right)}} \approx 1.91$

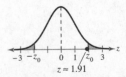

$z \approx 1.91$

**g.** Because $1.96 < z < 1.96$, fail to reject $H_0$.

**h.** There is not enough evidence at the 5% level of significance to support the claim that there is a difference between the proportion 40- to 49-year olds who are yoga users and the proportion of 40- to 49-year-olds who are non-yoga users.

**2a.** $\bar{p} = \dfrac{x_1 + x_2}{n_1 + n_2} = \dfrac{239 + 5990}{1593 + 29948} \approx 0.1975, \ \bar{q} = 1 - \bar{p} \approx 0.8025$

**b.** $n_1\bar{p} = 1593(0.1975) \approx 314.6 > 5$, $n_1\bar{q} = 1593(0.8025) \approx 1278.4 > 5$

$n_2\bar{p} = 29,948(0.1975) \approx 5914.7 > 5$, $n_2\bar{q} = 29,948(0.8025) \approx 24,033.3 > 5$

**c.** The claim is "the proportion of yoga users with incomes of \$20,000 to \$34,499 is less than the proportion of non-yoga users with incomes of \$20,000 to \$34,499."

$H_0 : p_1 \geq p_2$; $H_a : p_1 < p_2$ (claim)

**d.** $\alpha = 0.05$

**e.** $z_0 = -1.645$; Rejection region: $z < -1.645$

**f.** $z = \dfrac{(\hat{p}_1 - \hat{p}_2) - (p_1 - p_2)}{\sqrt{\bar{p}\bar{q}\left(\dfrac{1}{n_1} + \dfrac{1}{n_2}\right)}} = \dfrac{(0.15 - 0.20) - (0)}{\sqrt{0.1975 \cdot 0.8025\left(\dfrac{1}{1593} + \dfrac{1}{29,948}\right)}} \approx -4.88$

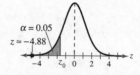

**g.** Because $z < -1.645$, reject $H_0$.

**h.** There is enough evidence at the 5% level of significance to support the claim that the proportion of yoga users with incomes of \$20,000 to \$34,499 is less than the proportion of non-yoga users with incomes of \$20,000 to \$34,499.

## 8.4 EXERCISE SOLUTIONS

**1.** (1) The samples are randomly selected.

(2) The samples are independent.

(3) $n_1\bar{p} \geq 5$, $n_1\bar{q} \geq 5$, $n_2\bar{p} \geq 5$, $n_2\bar{q} \geq 5$

**3.** $\bar{p} = \dfrac{x_1 + x_2}{n_1 + n_2} = \dfrac{35 + 36}{70 + 60} \approx 0.5462$; $\bar{q} = 1 - \bar{p} \approx 0.4538$

$n_1\bar{p} = 70(0.5462) \approx 38.2 > 5$, $n_1\bar{q} = 70(0.4538) \approx 31.8 > 5$

$n_2\bar{p} = 60(0.5462) \approx 32.8 > 5$, $n_2\bar{q} = 60(0.4538) \approx 27.2 > 5$

Because all conditions are met above, the normal sampling distribution can be used.

$H_0 : p_1 = p_2$; $H_a : p_1 \neq p_2$ (claim)

$-z_0 = -2.575$, $z_0 = 2.575$; Rejection regions: $z < -2.575$, $z > 2.575$

$z = \dfrac{(\hat{p}_1 - \hat{p}_2) - (p_1 - p_2)}{\sqrt{\bar{p}\bar{q}\left(\dfrac{1}{n_1} + \dfrac{1}{n_2}\right)}} = \dfrac{(0.5 - 0.6) - (0)}{\sqrt{0.5462 \cdot 0.4538\left(\dfrac{1}{70} + \dfrac{1}{60}\right)}} \approx -1.142$

Because $-2.575 < z < 2.575$, fail to reject $H_0$.

**5.** $\bar{p} = \dfrac{x_1 + x_2}{n_1 + n_2} = \dfrac{42 + 76}{150 + 200} \approx 0.337, \ \bar{q} = 1 - \bar{p} \approx 0.663$

$n_1\bar{p} \approx 150(0.337) \approx 50.57 > 5, \ n_1\bar{q} \approx 150(0.663) \approx 99.44 > 5.$

$n_2\bar{p} \approx 200(0.337) \approx 67.42 > 5, \text{ and } n_2\bar{q} \approx 200(0.663) \approx 132.58 > 5$

Because all conditions are met above, the normal sampling distribution can be used.

$H_0 : p_1 = p_2$ (claim); $H_a : p_1 \neq p_2$

$-z_0 = -1.645, \ z_0 = 1.645$; Rejection regions: $z < -1.645, \ z > 1.645$

$z = \dfrac{(\hat{p}_1 - \hat{p}_2) - (p_1 - p_2)}{\sqrt{\bar{p}\bar{q}\left(\dfrac{1}{n_1} + \dfrac{1}{n_2}\right)}} = \dfrac{(0.28 - 0.38) - (0)}{\sqrt{(0.337)(0.663)\left(\dfrac{1}{150} + \dfrac{1}{200}\right)}} \approx -1.96$

Because $z < -1.645$, reject $H_0$.

**7. (a)** The claim is "there is a difference in the proportion of subjects who feel all or mostly better after 4 weeks between subjects who used magnetic insoles and subjects who used nonmagnetic insoles."

$H_0 : p_1 = p_2; \ H_a : p_1 \neq p_2$ (claim)

**(b)** $-z_0 = -2.575, \ z_0 = 2.575$; Rejection regions: $z < -2.575, \ z > 2.575$

**(c)** $\bar{p} = \dfrac{x_1 + x_2}{n_1 + n_2} = \dfrac{17 + 18}{54 + 41} \approx 0.368, \ \bar{q} = 1 - \bar{p} \approx 0.632$

$z = \dfrac{(\hat{p}_1 - \hat{p}_2) - (p_1 - p_2)}{\sqrt{\bar{p}\bar{q}\left(\dfrac{1}{n_1} + \dfrac{1}{n_2}\right)}} = \dfrac{(0.315 - 0.439) - (0)}{\sqrt{(0.368)(0.632)\left(\dfrac{1}{54} + \dfrac{1}{41}\right)}} \approx -1.24$

**(d)** Because $-2.575 < z < 2.575$, fail to reject $H_0$.

**(e)** There is not enough evidence at the 1% level of significance to support the claim that there is a difference in the proportion of subjects who feel all or mostly better after 4 weeks between subjects who used magnetic insoles and subjects who used nonmagnetic insoles.

**9. (a)** The claim is "the proportion of males ages 18 to 24 who enrolled in college is less than the proportion of females ages 18 to 24 who enrolled in college."

$H_0 : p_1 \geq p_2; \ H_a : p_1 < p_2$ (claim)

**(b)** $z_0 = -1.645$; Rejection region: $z < -1.645$

**(c)** $\bar{p} = \dfrac{x_1 + x_2}{n_1 + n_2} = \dfrac{78 + 99}{200 + 220} \approx 0.421, \ \bar{q} = 1 - \bar{p} \approx 0.579$

$z = \dfrac{(\hat{p}_1 - \hat{p}_2) - (p_1 - p_2)}{\sqrt{\bar{p}\bar{q}\left(\dfrac{1}{n_1} + \dfrac{1}{n_2}\right)}} = \dfrac{(0.39 - 0.45) - (0)}{\sqrt{(0.421)(0.579)\left(\dfrac{1}{200} + \dfrac{1}{220}\right)}} \approx -1.24$

**(d)** Because $z > -1.645$, fail to reject $H_0$.

(e) There is not enough evidence at the 5% level of significance to support the claim that the proportion of males who enrolled in college is less than the proportion of females who enrolled in college.

11. (a) The claim is "the proportion of drivers who wear seat belts is greater in the South than in the Northeast."

$H_0 : p_1 \leq p_2; \ H_a : p_1 > p_2$ (claim)

(b) $z_0 = 1.645$; Rejection region: $z > 1.645$

(c) $\bar{p} = \dfrac{x_1 + x_2}{n_1 + n_2} = \dfrac{408 + 2881}{480 + 360} \approx 0.829, \ \bar{q} = 1 - \bar{p} \approx 0.171$

$z = \dfrac{(\hat{p}_1 - \hat{p}_2) - (p_1 - p_2)}{\sqrt{\bar{p}\bar{q}\left(\dfrac{1}{n_1} + \dfrac{1}{n_2}\right)}} = \dfrac{(0.85 - 0.80) - (0)}{\sqrt{(0.829)(0.171)\left(\dfrac{1}{480} + \dfrac{1}{360}\right)}} \approx 1.90$

(d) Because $z > 1.645$, reject $H_0$.

(e) There is enough evidence at the 5% level of significance to support the claim that the proportion of drivers who wear seat belts is greater in the South than in the Northeast.

13. The claim is "the proportion of adults in the United States who favor building new nuclear power plants in their country is the same as the proportion of adults from Great Britain who favor building new nuclear power plants in their country."

$H_0 : p_1 = p_2$ (claim); $H_a : p_1 \neq p_2$

$-z_0 = -1.96, \ z_0 = 1.96$; Rejection regions: $z < -1.96, \ z > 1.96$

$\bar{p} = \dfrac{x_1 + x_2}{n_1 + n_2} = \dfrac{404 + 317}{600 + 500} \approx 0.655, \ \bar{q} = 1 - \bar{p} \approx 0.345$

$z = \dfrac{(\hat{p}_1 - \hat{p}_2) - (p_1 - p_2)}{\sqrt{\bar{p}\bar{q}\left(\dfrac{1}{n_1} + \dfrac{1}{n_2}\right)}} = \dfrac{(0.51 - 0.50) - (0)}{\sqrt{(0.505)(0.495)\left(\dfrac{1}{1002} + \dfrac{1}{1056}\right)}} \approx 0.454$

Because $-1.96 < z < 1.96$, fail to reject $H_0$.

There is not enough evidence at the 5% level of significance to reject the claim that the proportion of adults in the United States who favor building new nuclear power plants in their country is the same as the proportion of adults from Great Britain who favor building new nuclear power plants in their country.

15. The claim is "the proportion of adults in France who favor building new nuclear power plants in their country is greater than the proportion of adults in Spain who favor building new nuclear power plants in their country."

$H_0 : p_1 \leq p_2; \ H_a : p_1 > p_2$ (claim)

$z_0 = 2.33$; Rejection region: $z > 2.33$

$\bar{p} = \dfrac{x_1 + x_2}{n_1 + n_2} = \dfrac{529 + 372}{1102 + 1006} \approx 0.427, \ \bar{q} = 1 - \bar{p} \approx 0.573$

$z = \dfrac{(\hat{p}_1 - \hat{p}_2) - (p_1 - p_2)}{\sqrt{\bar{p}\bar{q}\left(\dfrac{1}{n_1} + \dfrac{1}{n_2}\right)}} = \dfrac{(0.48 - 0.37) - (0)}{\sqrt{(0.427)(0.573)\left(\dfrac{1}{1102} + \dfrac{1}{1006}\right)}} \approx 5.100$

Because $z > 2.33$, reject $H_0$.

There is enough evidence at the 1% level of significance to support the claim that the proportion of adults in France who favor building new nuclear power plants in their country is greater than the proportion of adults in Spain who favor building new nuclear power plants in their country.

17. The claim is "the proportion of men ages 18 to 24 living in their parents' homes was greater in 2012 than in 2000."

$H_0 : p_1 \geq p_2; H_a : p_1 < p_2$ (claim)

$z_0 = -1.645$; Rejection region: $z < -1.645$

$$\bar{p} = \frac{x_1 + x_2}{n_1 + n_2} = \frac{141 + 156}{250 + 260} \approx 0.582, \quad \bar{q} = 1 - \bar{p} \approx 0.418$$

$$z = \frac{(\hat{p}_1 - \hat{p}_2) - (p_1 - p_2)}{\sqrt{\bar{p}\bar{q}\left(\dfrac{1}{n_1} + \dfrac{1}{n_2}\right)}} = \frac{(0.564 - 0.60) - (0)}{\sqrt{(0.582)(0.418)\left(\dfrac{1}{250} + \dfrac{1}{260}\right)}} \approx -0.824$$

Because $z > -1.645$, fail to reject $H_0$.

There is not enough evidence at the 5% level of significance to support the claim that the proportion of men ages 18 to 24 living in their parents' homes was greater in 2012 than in 2000.

19. The claim is "the proportion of 18- to 24-year-olds living in their parents' homes in 2000 was the same for men and women."

$H_0 : p_1 = p_2$ (claim); $H_a : p_1 \neq p_2$

$-z_0 = -2.575, z_0 = 2.575$; Rejection regions: $z < -2.575, z > 2.575$

$$\bar{p} = \frac{x_1 + x_2}{n_1 + n_2} = \frac{141 + 119}{250 + 280} \approx 0.491, \quad \bar{q} = 1 - \bar{p} \approx 0.509$$

$$z = \frac{(\hat{p}_1 - \hat{p}_2) - (p_1 - p_2)}{\sqrt{\bar{p}\bar{q}\left(\dfrac{1}{n_1} + \dfrac{1}{n_2}\right)}} = \frac{(0.564 - 0.425) - (0)}{\sqrt{(0.491)(0.509)\left(\dfrac{1}{250} + \dfrac{1}{280}\right)}} \approx 3.195$$

Because $z > 2.575$, reject $H_0$.

There is enough evidence at the 1% level of significance to reject the claim that the proportion of 18- to 24-year-olds living in their parents' homes in 2000 was the same for men and women.

21.

$$(\hat{p}_1 - \hat{p}_2) - z_c \sqrt{\frac{\hat{p}_1\hat{q}_1}{n_1} + \frac{\hat{p}_2\hat{q}_2}{n_2}} < p_1 - p_2 < (\hat{p}_1 - \hat{p}_2) + z_c \sqrt{\frac{\hat{p}_1\hat{q}_1}{n_1} + \frac{\hat{p}_2\hat{q}_2}{n_2}}$$

$$(0.06 - 0.09) - 1.96\sqrt{\frac{(0.06)(0.94)}{10,000} + \frac{(0.09)(0.91)}{8000}} < p_1 - p_2 < (0.06 - 0.09) + 1.96\sqrt{\frac{(0.06)(0.94)}{10,000} + \frac{(0.09)(0.91)}{8000}}$$

$$-0.03 - 0.008 < p_1 - p_2 < -0.03 + 0.008$$

$$-0.038 < p_1 - p_2 < -0.022$$

# CHAPTER 8 REVIEW EXERCISE SOLUTIONS

1. Dependent because the same adults were sampled.

3. Independent because different vehicles were sampled.

5. $H_0 : \mu_1 \geq \mu_2$ (claim); $H_a : \mu_1 < \mu_2$

   $z_0 = -1.645$; Rejection region: $z < -1.645$

   $$z = \frac{(\bar{x}_1 - \bar{x}_2) - (\mu_1 - \mu_2)}{\sqrt{\dfrac{\sigma_1^2}{n_1} + \dfrac{\sigma_2^2}{n_2}}} = \frac{(1.28 - 1.34) - (0)}{\sqrt{\dfrac{(0.30)^2}{96} + \dfrac{(0.23)^2}{85}}} \approx -1.519$$

   Because $z > -1.645$, fail to reject $H_0$. There is not enough evidence at the 5% level of significance to support the claim.

7. $H_0 : \mu_1 \geq \mu_2$; $H_a : \mu_1 < \mu_2$ (claim)

   $z_0 = -1.28$; Rejection regions: $z < -1.28$

   $$z = \frac{(\bar{x}_1 - \bar{x}_2) - (\mu_1 - \mu_2)}{\sqrt{\dfrac{\sigma_1^2}{n_1} + \dfrac{\sigma_2^2}{n_2}}} = \frac{(0.28 - 0.33) - (0)}{\sqrt{\dfrac{(0.11)^2}{41} + \dfrac{(0.10)^2}{34}}} \approx -2.060$$

   Because $z < -1.28$, reject $H_0$. There is enough evidence at the 10% level of significance to support the claim.

9. (a) The claim is "the mean sodium content of chicken sandwiches at Restaurant A is less than the mean sodium content of chicken sandwiches at Restaurant B."
   $H_0 : \mu_1 \geq \mu_2$; $H_a : \mu_1 < \mu_2$ (claim)
   (b) $z_0 = -1.645$; Rejection region: $z < -1.645$
   (c) $z = \dfrac{(\bar{x}_1 - \bar{x}_2) - (\mu_1 - \mu_2)}{\sqrt{\dfrac{\sigma_1^2}{n_1} + \dfrac{\sigma_2^2}{n_2}}} = \dfrac{(670 - 690) - (0)}{\sqrt{\dfrac{(20)^2}{22} + \dfrac{(30)^2}{28}}} \approx -2.82$
   (d) Because $z < -1.645$, reject $H_0$.
   (e) There is enough evidence at the 5% level of significance to support the claim that the mean sodium content of chicken sandwiches at Restaurant A is less than the mean sodium content of chicken sandwiches at Restaurant B.

11. $H_0 : \mu_1 = \mu_2$ (claim); $H_a : \mu_1 \neq \mu_2$

    d.f. $= n_1 + n_2 - 2 = 31$

    $-t_0 = -2.040$, $t_0 = 2.040$; Rejection regions: $t < -2.040$, $t > 2.040$

$$t = \frac{(\bar{x}_1 - \bar{x}_2) - (\mu_1 - \mu_2)}{\sqrt{\frac{(n_1 - 1)s_1^2 + (n_2 - 1)s_2^2}{n_1 + n_2 - 2}} \sqrt{\frac{1}{n_1} + \frac{1}{n_2}}} = \frac{(228 - 207) - (0)}{\sqrt{\frac{(20 - 1)(27)^2 + (13 - 1)(25)^2}{20 + 13 - 2}} \sqrt{\frac{1}{20} + \frac{1}{13}}} \approx 2.25$$

Because $t > 2.040$, reject $H_0$. There is enough evidence at the 5% level of significance to reject the claim.

13. $H_0 : \mu_1 \le \mu_2$ (claim); $H_a : \mu_1 > \mu_2$

d.f. $= \min\{n_1 - 1, \ n_2 - 1\} = 24$

$t_0 = 1.711$; Rejection region: $t > 1.711$

$$t = \frac{(\bar{x}_1 - \bar{x}_2) - (\mu_1 - \mu_2)}{\sqrt{\frac{s_1^2}{n_1} + \frac{s_2^2}{n_2}}} = \frac{(183.5 - 184.7) - (0)}{\sqrt{\frac{(1.3)^2}{25} + \frac{(3.9)^2}{25}}} \approx -1.460$$

Because $t < 1.711$, fail to reject $H_0$. There is not enough evidence at the 5% level of significance to reject the claim.

15. $H_0 : \mu_1 = \mu_2$; $H_a : \mu_1 \neq \mu_2$ (claim)

d.f. $= n_1 + n_2 - 2 = 10$

$-t_0 = -3.169$, $t_0 = 3.169$; Rejection regions: $t < -3.169$, $t > 3.169$

$$t = \frac{(\bar{x}_1 - \bar{x}_2) - (\mu_1 - \mu_2)}{\sqrt{\frac{(n - 1)s_1^2 + (n_2 - 1)s_2^2}{n_1 + n_2 - 2}} \cdot \sqrt{\frac{1}{n_1} + \frac{1}{n_2}}} = \frac{(61 - 55) - (0)}{\sqrt{\frac{(5 - 1)(3.3)^2 + (7 - 1)(1.2)^2}{5 + 7 - 2}} \cdot \sqrt{\frac{1}{5} + \frac{1}{7}}} \approx 4.484$$

Because $t > 3.169$, reject $H_0$. There is enough evidence at the 1% level of significance to support the claim.

17. (a) The claim is "third graders taught with the directed reading activities scored higher than those taught without the activities."
$H_0 : \mu_1 \le \mu_2$; $H_a : \mu_1 > \mu_2$ (claim)

(b) d.f. $= n_1 + n_2 - 2 = 42$
$t_0 = 1.648$; Rejection region: $t > 1.648$

(c) $\bar{x}_1 \approx 51.476$, $s_1 \approx 11.007$, $n_1 = 21$
$\bar{x}_2 \approx 41.522$, $s_2 \approx 17.149$, $n_2 = 23$
$$t = \frac{(\bar{x}_1 - \bar{x}_2) - (\mu_1 - \mu_2)}{\sqrt{\frac{(n - 1)s_1^2 + (n_2 - 1)s_2^2}{n_1 + n_2 - 2}}} \approx \frac{(51.476 - 41.522) - (0)}{\sqrt{\frac{(21 - 1)(11.007)^2 + (23 - 1)(17.149)^2}{21 + 23 - 2}} \sqrt{\frac{1}{21} + \frac{1}{23}}} \approx 2.266$$

(d) Because $t > 1.648$, reject $H_0$.

(e) There is enough evidence at the 5% level of significance to support the claim that third graders taught with the directed reading activities scored higher than those taught without the activities.

**19.** $H_0 : \mu_d = 0$ (claim); $H_a : \mu_d \neq 0$

$\alpha = 0.01$ and d.f. $= n - 1 = 15$

$-t_0 = -2.947$, $t_0 = 2.947$; Rejection regions: $t < -2.947$, $t > 2.947$

$t = \dfrac{\overline{d} - \mu_d}{\dfrac{s_d}{\sqrt{n}}} = \dfrac{8.5 - 0}{\dfrac{10.7}{\sqrt{16}}} \approx 3.178$

Because $t > 2.947$, reject $H_0$. There is enough evidence at the 1% level of significance to reject the claim.

**21.** $H_0 : \mu_d \leq 0$ (claim); $H_a : \mu_d > 0$

$\alpha = 0.10$ and d.f. $= n - 1 = 33$

$t_0 = 1.308$; Rejection region: $t > 1.308$

$t = \dfrac{\overline{d} - \mu_d}{\dfrac{s_d}{\sqrt{n}}} = \dfrac{10.3 - 0}{\dfrac{18.19}{\sqrt{33}}} \approx 3.253$

Because $t > 1.308$, reject $H_0$. There is enough evidence at the 10% level of significance to reject the claim.

**23.** (a) The claim is "calcium supplements can decrease the systolic blood pressures of men."

   $H_0 : \mu_d \leq 0$; $H_a : \mu_d > 0$ (claim)

   (b) $t_0 = 1.383$; Rejection region: $t > 1.383$

   (c) $\overline{d} = 5$ and $s_d \approx 8.743$

   (d) $t = \dfrac{\overline{d} - \mu_d}{\dfrac{s_d}{\sqrt{n}}} = \dfrac{5 - 0}{\dfrac{8.743}{\sqrt{10}}} \approx 1.808$

   (e) Because $t > 1.383$, reject $H_0$.

   (f) There is enough evidence at the 10% level of significance to support the claim that calcium supplements can decrease the systolic blood pressures of men.

**25.** $\overline{p} = \dfrac{x_1 + x_2}{n_1 + n_2} = \dfrac{425 + 410}{840 + 760} \approx 0.522$, $\overline{q} = 1 - \overline{p} \approx 0.478$

$n_1 \overline{p} \approx 840(0.522) \approx 438.48 > 5$, $n_1 \overline{q} \approx 840(0.478) \approx 401.52 > 5$,

$n_2 \overline{p} \approx 760(0.522) \approx 396.72 > 5$, and $n_2 \overline{q} \approx 760(0.478) \approx 363.28 > 5$.

Can use normal sampling distribution.

$H_0 : p_1 = p_2$ (claim); $H_a : p_1 \neq p_2$

$-z_0 = -1.96$, $z_0 = 1.96$; Rejection regions: $z < -1.96$, $z > 1.96$

$z = \dfrac{(\hat{p}_1 - \hat{p}_2) - (p_1 - p_2)}{\sqrt{\overline{p}\,\overline{q}\left(\dfrac{1}{n_1} + \dfrac{1}{n_2}\right)}} = \dfrac{(0.506 - 0.539) - (0)}{\sqrt{0.522(0.478)\left(\dfrac{1}{840} + \dfrac{1}{760}\right)}} \approx -1.320$

Because $-1.96 < z < 1.96$, fail to reject $H_0$. There is not enough evidence at the 5% level of significance to reject the claim.

27. $\bar{p} = \dfrac{x_1 + x_2}{n_1 + n_2} = \dfrac{261 + 207}{556 + 483} \approx 0.450, \ \bar{q} = 1 - \bar{p} \approx 0.550$

$n_1\bar{p} \approx 556(0.450) \approx 250.2 > 5, \ n_1\bar{q} \approx 556(0.550) \approx 305.8 > 5,$
$n_2\bar{p} \approx 483(0.450) \approx 217.35 > 5, \ \text{and } n_2\bar{q} \approx 483(0.550) \approx 265.65 > 5 .$
Can use normal sampling distribution.

$H_0 : p_1 \le p_2; \ H_a : p_1 > p_2$ (claim)

$z_0 = 1.28;$ Rejection region: $z > 1.28$

$z = \dfrac{(\hat{p}_1 - \hat{p}_2) - (p_1 - p_2)}{\sqrt{\bar{p}\bar{q}\left(\dfrac{1}{n_1} + \dfrac{1}{n_2}\right)}} = \dfrac{(0.469 - 0.429) - (0)}{\sqrt{0.450(0.550)\left(\dfrac{1}{556} + \dfrac{1}{483}\right)}} \approx 1.293$

Because $z > 1.28$, reject $H_0$. There is enough evidence at the 10% level of significance to support the claim.

29. (a) The claim is "the proportion of subjects who are pain-free is the same for the two groups."
$H_0 : p_1 = p_2$ (claim); $H_a : p_1 \ne p_2$

(b) $-z_0 = -1.96, z_0 = 1.96;$ Rejection regions: $z < -1.96, z > 1.96$

(c) $\bar{p} = \dfrac{x_1 + x_2}{n_1 + n_2} = \dfrac{100 + 41}{400 + 400} \approx 0.176, \ \bar{q} = 1 - \bar{p} \approx 0.824$

$z = \dfrac{(\hat{p}_1 - \hat{p}_2) - (p_1 - p_2)}{\sqrt{\bar{p}\bar{q}\left(\dfrac{1}{n_1} + \dfrac{1}{n_2}\right)}} = \dfrac{(0.250 - 0.101) - (0)}{\sqrt{0.174(0.826)\left(\dfrac{1}{400} + \dfrac{1}{407}\right)}} \approx 5.58$

(d) Because $z > 1.96$, reject $H_0$.

(e) There is enough evidence at the 5% level of significance to reject the claim that the proportion of subjects who are pain-free is the same for the two groups.

## CHAPTER 8 QUIZ SOLUTIONS

1. (a) The claim is "the mean score on the science assessment test for male high school students is greater than the mean score for female high school students."
$H_0 : \mu_1 \le \mu_2; \ H_a : \mu_1 > \mu_2$ (claim)

(b) Right-tailed because $H_a$ contains $>$; $z$-test because $\sigma_1$ and $\sigma_2$ are known, the samples are random samples, the samples are independent, and $n_1 \ge 30$ and $n_2 \ge 30$.

(c) $z_0 = 1.645;$ Rejection region: $z > 1.645$

(d) $z = \dfrac{(\bar{x}_1 - \bar{x}_2) - (\mu_1 - \mu_2)}{\sqrt{\dfrac{\sigma_1^2}{n_1} + \dfrac{\sigma_2^2}{n_2}}} = \dfrac{(153 - 147) - (0)}{\sqrt{\dfrac{(36)^2}{49} + \dfrac{(34)^2}{50}}} \approx 0.852$

(e) Because $z < 1.645$, fail to reject $H_0$.

(f) There is not enough evidence at the 5% level of significance to support the claim that the mean score on the science assessment test for male high school students is greater than for the female high school students.

2. (a) The claim is "the mean scores on a science assessment test for fourth grade boys and girls are equal."

$H_0 : \mu_1 = \mu_2$ (claim); $H_a : \mu_1 \neq \mu_2$

(b) Two-tailed because $H_a$ contains $\neq$; $t$-test because $\sigma_1$ and $\sigma_2$ are unknown, the samples are random samples, the samples are independent, and the populations are normally distributed.

(c) d.f. $= n_1 + n_2 - 2 = 26$

$-t_0 = -2.779$, $t_0 = 2.779$; Rejection regions: $t < -2.779$, $t > 2.779$

(d) $t = \dfrac{(\bar{x}_1 - \bar{x}_2) - (\mu_1 - \mu_2)}{\sqrt{\dfrac{(n-1)s_1^2 + (n_2-1)s_2^2}{n_1 + n_2 - 2}} \cdot \sqrt{\dfrac{1}{n_1} + \dfrac{1}{n_2}}} = \dfrac{(151 - 149) - (0)}{\sqrt{\dfrac{(13-1)(36)^2 + (15-1)(34)^2}{13 + 15 - 2}} \cdot \sqrt{\dfrac{1}{13} + \dfrac{1}{15}}} \approx 0.151$

(e) Because $-2.779 < t < 2.779$, fail to reject $H_0$.

(f) There is not enough evidence at the 1% level of significance to reject the claim that the mean scores on the science assessment test are the same for fourth grade boys and girls.

3. (a) The claim is "the seminar helps adults increase their credit scores."

$H_0 : \mu_d \geq 0$; $H_a : \mu_d < 0$ (claim)

(b) Left-tailed because $H_a$ contains $<$; $t$-test because both populations are normally distributed and the samples are dependent.

(c) d.f. $= n - 1 = 11$

$t_0 = -2.718$; Rejection region: $t < -2.718$

(d) $t = \dfrac{\bar{d} - \mu_d}{\dfrac{s_d}{\sqrt{n}}} = \dfrac{-51.167 - 0}{\dfrac{34.938}{\sqrt{12}}} \approx -5.073$

(e) Because $t < -2.718$, reject $H_0$.

(f) There is enough evidence at the 1% level of significance to support the claim that the seminar helps adults increase their credit scores.

4. (a) The claim is "the proportion of U.S. adults who favor mandatory testing to assess how well schools are educating students is less than it was 9 years ago."

$H_0 : p_1 \geq p_2$; $H_a : p_1 < p_2$ (claim)

(b) Left-tailed because $H_a$ contains $<$; $z$-test because you are testing proportions, the samples are random samples, the samples are independent and the quantities $n_1\bar{p}$, $n_1\bar{q}$, $n_2\bar{p}$, and $n_2\bar{q}$ are at least 5.

(c) $z_0 = 1.645$; Rejection region: $z < -1.645$

(d) $\bar{p} = \dfrac{x_1 + x_2}{n_1 + n_2} = \dfrac{863 + 823}{1216 + 1002} \approx 0.760$, $\bar{q} = 1 - \bar{p} \approx 0.240$

$$z = \frac{(\hat{p}_1 - \hat{p}_2) - (p_1 - p_2)}{\sqrt{\overline{pq}\left(\dfrac{1}{n_1} + \dfrac{1}{n_2}\right)}} = \frac{(0.710 - 0.821) - (0)}{\sqrt{0.760(0.240)\left(\dfrac{1}{1216} + \dfrac{1}{1002}\right)}} \approx -6.09$$

(e) Because $z < -1.645$, reject $H_0$.

(f) There is enough evidence at the 5% level of significance to support the claim that the proportion of U.S. adults who favor mandatory testing to assess how well schools are educating students is less than it was 9 years ago.

## CUMULATIVE REVIEW, CHAPTERS 6–8

1. (a) $\hat{p} = 0.15$, $\hat{q} = 0.85$

$$\hat{p} \pm z_c \sqrt{\frac{\hat{p}\hat{q}}{n}} = 0.15 \pm 1.96 \sqrt{\frac{(0.15)(0.85)}{1000}} \approx 0.15 \pm 0.022 = (0.128,\ 0.172)$$

   (b) $H_0 : p \le 0.12$; $H_a : p > 0.12$ (claim)

$z_0 = 1.645$; Rejection region: $z > 1.645$

$$z = \frac{\hat{p} - p}{\sqrt{\dfrac{pq}{n}}} = \frac{0.15 - 0.12}{\sqrt{\dfrac{(0.12)(0.88)}{1000}}} \approx 2.92$$

Because $z > 1.645$, reject $H_0$.

There is enough evidence at the 5% level of significance to support the claim that more than 12% of people who attend community college are age 40 or older.

3. $\overline{x} \pm z_c \dfrac{\sigma}{\sqrt{n}} = 26.97 \pm 1.96 \dfrac{3.4}{\sqrt{42}} \approx 26.97 \pm 1.03 = (25.94,\ 28.00)$; $z$-distribution

5. $\overline{x} \pm t_c \dfrac{s}{\sqrt{n}} = 12.1 \pm 2.787 \dfrac{2.64}{\sqrt{26}} \approx 12.1 \pm 1.4 = (10.7,\ 13.5)$; $t$-distribution

7. $H_0 : \mu_1 \le \mu_2$; $H_a : \mu_1 > \mu_2$ (claim)

$z_0 = 1.28$; Rejection region: $z > 1.28$

$$z = \frac{(\overline{x}_1 - \overline{x}_2) - 0}{\sqrt{\dfrac{\sigma_1^2}{n_1} + \dfrac{\sigma_2^2}{n_2}}} = \frac{(3086 - 2263) - (0)}{\sqrt{\dfrac{(563)^2}{85} + \dfrac{(624)^2}{68}}} \approx 8.464$$

Because $z > 1.28$, reject $H_0$. There is enough evidence at the 10% level of significance to support the claim that the mean birth weight of a single-birth baby is greater than the mean birth weight of a baby that has a twin.

9. $H_0 : p \ge 0.19$ (claim)

$H_a : p < 0.19$

11. $H_0 : \mu = 2.28$

$H_a : \mu \neq 2.28$ (claim)

**13.** $H_0 : \mu_1 = \mu_2;\ H_a : \mu_1 \neq \mu_2$ (claim)

d.f. $= n_1 + n_2 - 2 = 26 + 18 - 2 = 42$

$-t_0 = -2.021,\ t_0 = 2.021$; Rejection regions: $t < -2.021,\ t > 2.021$

$$t = \frac{(\bar{x}_1 - \bar{x}_2) - (0)}{\sqrt{\dfrac{(n_1 - 1)s_1^2 + (n_2 - 1)s_2^2}{n_1 + n_2 - 2}}\sqrt{\dfrac{1}{n_1} + \dfrac{1}{n_2}}} = \frac{1783 - 2064}{\sqrt{\dfrac{(26 - 1)(218)^2 + (18 - 1)(186)^2}{26 + 18 - 2}}\sqrt{\dfrac{1}{26} + \dfrac{1}{18}}} \approx -4.456$$

Because $t < -2.021$, reject $H_0$. There is enough evidence at the 5% level of significance to support the claim that the mean SAT scores for male athletes and male non-athletes at a college are different.

**15.** $H_0 : p_1 = p_2$ (claim); $H_a : p_1 \neq p_2$

$$\bar{p} = \frac{x_1 + x_2}{n_1 + n_2} = \frac{195 + 204}{319 + 323} \approx 0.622,\ \bar{q} = 1 - \bar{p} \approx 0.378$$

$-z_0 = -1.645,\ z_0 = 1.645$; Rejection regions: $z < -1.645,\ z > 1.645$;

$$z_0 = \frac{(\hat{p}_1 - \hat{p}_2) - 0}{\sqrt{\bar{p}\bar{q}\left(\dfrac{1}{n_1} + \dfrac{1}{n_2}\right)}} = \frac{(0.611 - 0.632)}{\sqrt{(0.622)(0.378)\left(\dfrac{1}{319} + \dfrac{1}{323}\right)}} \approx -0.549$$

Because $-1.645 < z < 1.645$;, fail to reject $H_0$. There is not enough evidence at the 10% level of significance to reject the claim that the proportions of players sustaining head and neck injuries are the same for the two groups.

**17.** A type I error will occur when the actual proportion of dog owners that dress their dogs in outfits is 0.18 but you reject $H_0$. A type II error will occur when the actual proportion of dog owners who dress their dogs in outfits is different from 0.18 but you fail to reject $H_0$.

# Correlation and Regression

## 9.1 CORRELATION

## 9.1 Try It Yourself Solutions

**1ab.**

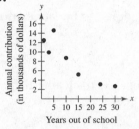

c. Yes, it appears that there is a negative linear correlation. As the number of years out of school increases, the annual contribution tends to decreases.

**2ab.**

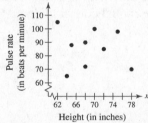

c. No, it appears that there is no linear correlation between height and pulse rate.

**3ab.**

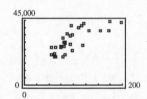

c. Yes, it appears that there is a positive linear correlation. As the team salary increases, the average attendance per home game tends to increases.

**4a.** $n = 7$

| $x$ | $y$ | $xy$ | $x^2$ | $y^2$ |
|---|---|---|---|---|
| 1 | 12.5 | 12.5 | 1 | 156.25 |
| 10 | 8.7 | 87.0 | 100 | 75.69 |
| 5 | 14.6 | 73.0 | 25 | 213.16 |
| 15 | 5.2 | 78.0 | 225 | 27.04 |
| 3 | 9.9 | 29.7 | 9 | 98.01 |
| 24 | 3.1 | 74.4 | 576 | 9.61 |
| 30 | 2.7 | 81.0 | 900 | 7.29 |
| $\sum x = 88$ | $\sum y = 56.7$ | $\sum xy = 435.6$ | $\sum x^2 = 1836$ | $\sum y^2 = 587.05$ |

**b.** $r = \dfrac{n\sum xy - (\sum x)(\sum y)}{\sqrt{n\sum x^2 - (\sum x)^2}\sqrt{n\sum y^2 - (\sum y)^2}}$

$= \dfrac{(7)(435.6) - (88)(56.7)}{\sqrt{(7)(1836) - (88)^2}\sqrt{(7)(587.05) - (56.7)^2}}$

$= \dfrac{-1940.4}{\sqrt{5108}\sqrt{894.46}} \approx -0.908$

**c.** Because $r$ is close to $-1$, this suggests a strong negative linear correlation. As the number of years out of school increases, the annual contribution tends to decrease.

**5ab.** $r \approx 0.769$

**c.** Because $r$ is close to 1, this suggests a strong positive linear correlation. As the team salaries increase, the average attendances at home games tends to increase.

**6a.** $n = 7$

**b.** $\alpha = 0.01$

**c.** 0.875

**d.** $|r| \approx |-0.908| > 0.875$; The correlation is significant.

**e.** There is enough evidence at the 1% level of significance to conclude that there is a significant linear correlation between the number of years out of school and the annual contribution.

**7a.** $H_0: \rho = 0; H_a: \rho \neq 0$

**b.** $\alpha = 0.01$

**c.** d.f. $= n - 2 = 28$

**d.** $-t_0 = -2.763, t_0 = 2.763$; Rejection regions: $t < -2.763$ or $t > 2.763$

**e.** $t = \dfrac{r}{\sqrt{\dfrac{1 - r^2}{n - 2}}} = \dfrac{0.769}{\sqrt{\dfrac{1 - (0.769)^2}{30 - 2}}} \approx 6.366$

**f.** Because $t > 2.763$, reject $H_0$.

**g.** There is enough evidence at the 1% level of significance to conclude that there is a significant linear correlation between the salaries and the average attendances at home games for the teams in Major League Baseball.

## 9.1 EXERCISE SOLUTIONS

1. Increase

3. The range of values for the correlation coefficient is $-1$ to 1, inclusive.

5. Answers will vary. *Sample answer*:
   Perfect positive linear correlation: price per gallon of gasoline and total cost of gasoline
   Perfect negative linear correlation: distance from door and height of wheelchair ramp

7. $r$ is the sample correlation coefficient, while $\rho$ is the population correlation coefficient.

**9.** Strong negative linear correlation

**11.** Perfect negative linear correlation

**13.** Strong positive linear correlation

**15.** (c), You would expect a positive linear correlation between age and income.

**17.** (b), You would expect a negative linear correlation between age and balance on student loans.

**19.** Explanatory variable: Amount of water consumed
Response variable: Weight loss

**21.** (a)

(b)

| $x$ | $y$ | $xy$ | $x^2$ | $y^2$ |
|---|---|---|---|---|
| 16 | 109 | 1744 | 256 | 11,881 |
| 25 | 122 | 3050 | 625 | 14,884 |
| 39 | 143 | 5577 | 1521 | 20,449 |
| 45 | 132 | 5940 | 2025 | 17,424 |
| 49 | 199 | 9751 | 2401 | 39,601 |
| 64 | 185 | 11,840 | 4096 | 34,225 |
| 70 | 199 | 13,930 | 4900 | 39,601 |
| 29 | 130 | 3770 | 841 | 16,900 |
| 57 | 175 | 9975 | 3249 | 30,625 |
| 22 | 118 | 2596 | 484 | 13,924 |
| $\sum x = 416$ | $\sum y = 1512$ | $\sum xy = 68,173$ | $\sum x^2 = 20,398$ | $\sum y^2 = 239,514$ |

$$r = \frac{n\sum xy - (\sum x)(\sum y)}{\sqrt{n\sum x^2 - (\sum x)^2}\sqrt{n\sum y^2 - (\sum y)^2}}$$

$$= \frac{10(68,173) - (416)(1512)}{\sqrt{10(20,398) - (416)^2}\sqrt{10(239,514) - (1512)^2}}$$

$$= \frac{52,738}{\sqrt{30,924}\sqrt{108,996}} \approx 0.908$$

(c) Strong positive linear correlation; As age increases, the systolic blood pressure tends to increase.

**23.** (a)

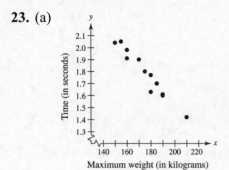

(b)

| x | y | xy | $x^2$ | $y^2$ |
|---|---|-----|-------|-------|
| 175 | 1.80 | 315 | 30,625 | 3.24 |
| 180 | 1.77 | 318.6 | 32,400 | 3.1329 |
| 155 | 2.05 | 317.75 | 24,025 | 4.2025 |
| 210 | 1.42 | 298.2 | 44,100 | 2.0164 |
| 150 | 2.04 | 306 | 22,500 | 4.1616 |
| 190 | 1.61 | 305.9 | 36,100 | 2.5921 |
| 185 | 1.70 | 314.5 | 34,225 | 2.89 |
| 160 | 1.91 | 305.6 | 25,600 | 3.6481 |
| 190 | 1.60 | 304 | 36,100 | 2.56 |
| 180 | 1.63 | 293.4 | 32,400 | 2.6569 |
| 160 | 1.98 | 316.8 | 25,600 | 3.9204 |
| 170 | 1.90 | 323 | 28,900 | 3.61 |
| $\sum x = 2105$ | $\sum y = 21.41$ | $\sum xy = 3718.75$ | $\sum x^2 = 372,575$ | $\sum y^2 = 38.6309$ |

$$r = \frac{n\sum xy - (\sum x)(\sum y)}{\sqrt{n\sum x^2 - (\sum x)^2}\sqrt{n\sum y^2 - (\sum y)^2}}$$

$$= \frac{12(3718.75) - (2105)(21.41)}{\sqrt{12(372,575) - (2105)^2}\sqrt{12(38.6309) - (21.41)^2}}$$

$$= \frac{-443.05}{\sqrt{39,875}\sqrt{5.1827}} \approx -0.975$$

(c) Strong negative linear correlation; As maximum weight for one repetition of a half squat increases, the time to run 10-meter sprint tends to decrease.

**25.** (a)

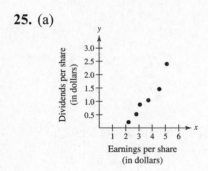

(b)

| x | y | xy | $x^2$ | $y^2$ |
|---|---|---|---|---|
| 2.79 | 0.52 | 1.4508 | 7.7841 | 0.2704 |
| 5.10 | 2.40 | 12.24 | 26.01 | 5.76 |
| 4.53 | 1.46 | 6.6138 | 20.5209 | 2.1316 |
| 3.06 | 0.88 | 2.6928 | 9.3636 | 0.7744 |
| 3.70 | 1.04 | 3.848 | 13.69 | 1.0816 |
| 2.20 | 0.22 | 0.484 | 4.84 | 0.0484 |
| $\sum x = 21.38$ | $\sum y = 6.52$ | $\sum xy = 27.3294$ | $\sum x^2 = 82.2086$ | $\sum y^2 = 10.0664$ |

$$r = \frac{n\sum xy - (\sum x)(\sum y)}{\sqrt{n\sum x^2 - (\sum x)^2}\sqrt{n\sum y^2 - (\sum y)^2}}$$

$$= \frac{6(27.3294) - (21.38)(6.52)}{\sqrt{6(82.2086) - (21.38)^2}\sqrt{6(10.0664) - (6.52)^2}}$$

$$= \frac{24.5788}{\sqrt{36.1472}\sqrt{17.888}} \approx 0.967$$

(c) Strong positive linear correlation; As the earnings per share increase, the dividends per share tend to increase.

**27.** The correlation coefficient gets stronger, going from $r \approx 0.908$ to $r \approx 0.969$.

**29.** The correlation coefficient gets weaker, going from $r \approx -0.975$ to $r \approx -0.655$.

**31.** $r \approx 0.623$

$n = 8$ and $\alpha = 0.01$

critical value $= 0.834$

$|r| \approx 0.623 < 0.834 \Rightarrow$ The correlation is not significant.

OR

$H_0: \rho = 0; H_a: \rho \neq 0$

$\alpha = 0.01$

d.f. $= n - 2 = 6$

$-t_0 = -3.707, t_0 = 3.707$; Rejection regions: $t < -3.707$ or $t > 3.707$.

$$t = \frac{r}{\sqrt{\dfrac{1 - r^2}{n - 2}}} = \frac{0.623}{\sqrt{\dfrac{1 - (0.623)^2}{8 - 2}}} \approx 1.951$$

Because $-3.707 < t < 3.707$, fail to reject $H_0$. There is not enough evidence at the 1% level of significance to conclude that there is a significant linear correlation between vehicle weight and the variability in braking distance on a dry surface.

**33.** $r \approx -0.975$

$n = 12$ and $\alpha = 0.01$

critical value $= 0.708$

$|r| = |-0.975| > 0.708 \Rightarrow$ The correlation is significant.

OR

$H_0 : \rho = 0; H_a : \rho \neq 0$

$\alpha = 0.01$

d.f. $= n - 2 = 10$

$-t_0 = -3.169$, $t_0 = 3.169$; Rejection regions: $t < -3.169$ or $t > 3.169$.

$$t = \frac{r}{\sqrt{\frac{1-r^2}{n-2}}} = \frac{-0.975}{\sqrt{\frac{1-(-0.975)^2}{12-2}}} \approx -13.876$$

Because $t < -3.169$, reject $H_0$. There is enough evidence at the 1% level of significance to conclude that there is a significant linear correlation between the maximum weight for one repetition of a half squat and the time to run a 10-meter sprint.

**35.**

| $x$ | $y$ | $xy$ | $x^2$ | $y^2$ |
|---|---|---|---|---|
| 1.80 | 175 | 315 | 3.24 | 30,625 |
| 1.77 | 180 | 318.6 | 3.1329 | 32,400 |
| 2.05 | 155 | 317.75 | 4.2025 | 24,025 |
| 1.42 | 210 | 298.2 | 2.0164 | 44,100 |
| 2.04 | 150 | 306 | 4.1616 | 22,500 |
| 1.61 | 190 | 305.9 | 2.5921 | 36,100 |
| 1.70 | 185 | 314.5 | 2.89 | 34,225 |
| 1.91 | 160 | 305.6 | 3.6481 | 25,600 |
| 1.60 | 190 | 304 | 2.56 | 36,100 |
| 1.63 | 180 | 293.4 | 2.6569 | 32,400 |
| 1.98 | 160 | 316.8 | 3.9204 | 25,600 |
| 1.90 | 170 | 323 | 3.61 | 28,900 |
| $\sum x = 21.41$ | $\sum y = 2105$ | $\sum xy = 3718.75$ | $\sum x^2 = 38.6309$ | $\sum y^2 = 372{,}575$ |

$$r = \frac{n\sum xy - (\sum x)(\sum y)}{\sqrt{n\sum x^2 - (\sum x)^2}\sqrt{n\sum y^2 - (\sum y)^2}}$$

$$= \frac{12(3718.75) - (21.41)(2105)}{\sqrt{12(38.6309) - (21.41)^2}\sqrt{12(372{,}575) - (2105)^2}}$$

$$= \frac{-443.05}{\sqrt{5.1827}\sqrt{39{,}875}} \approx -0.975$$

The correlation coefficient remains unchanged when the x-values and the y-values are switched.

## 9.2 LINEAR REGRESSION

### 9.2 Try It Yourself Solutions

**1a.** $n = 7, \sum x = 88, \sum y = 56.7, \sum xy = 435.6, \sum x^2 = 1836$

**b.** $m = \dfrac{n\sum xy - (\sum x)(\sum y)}{n\sum x^2 - (\sum x)^2} = \dfrac{(7)(435.6) - (88)(56.7)}{(7)(1836) - (88)^2} = \dfrac{-1940.4}{5108} \approx -0.379875$

$b = \bar{y} - m\bar{x} = \dfrac{\sum y}{n} - m\dfrac{\sum x}{n} \approx \dfrac{(56.7)}{7} - (-0.379875)\dfrac{(88)}{7} \approx 12.8756$

**c.** $\hat{y} = -0.380x + 12.876$

**2a.** Enter the data.

**b.** $m \approx 164.621, \ b \approx 14{,}746.961$

**c.** $\hat{y} = 164.621x + 14{,}746.961$

**3a.** (1) $\hat{y} = 12.481(2) + 33.683$   (2) $\hat{y} = 12.481(3.32) + 33.683$

**b.** (1) $\hat{y} = 58.645$   (2) $\hat{y} = 75.120$

**c.** (1) 58.645 minutes   (2) 75.120 minutes

### 9.2 EXERCISE SOLUTIONS

1.  A residual is the difference between the observed $y$-value of a data point and the predicted $y$-value on the regression line for the $x$-coordinate of the data point. A residual is positive when the data point is above the line, negative when the point is below the line, and zero when the observed $y$-value equals the predicted $y$-value.

3.  Substitute a value of $x$ into the equation of a regression line and solve for $\hat{y}$.

5.  The correlation between variables must be significant.

7.  b   **9.** e   **11.** F   **13.** c   **15.** a

**17.**

| $x$ | $y$ | $xy$ | $x^2$ |
|-----|-----|------|-------|
| 869 | 60 | 52,140 | 755,161 |
| 820 | 50 | 41,000 | 672,400 |
| 771 | 50 | 38,550 | 594,441 |
| 696 | 52 | 36,192 | 484,416 |
| 692 | 40 | 27,680 | 478,864 |
| 676 | 47 | 31,772 | 456,976 |
| 656 | 41 | 26,896 | 430,336 |
| 492 | 39 | 19,188 | 242,064 |
| 486 | 26 | 12,636 | 236,196 |
| $\sum x = 6158$ | $\sum y = 405$ | $\sum xy = 286,054$ | $\sum x^2 = 4,350,854$ |

$$m = \frac{n\sum xy - (\sum x)(\sum y)}{n\sum x^2 - (\sum x)^2} = \frac{(9)(286,054) - (6158)(405)}{(9)(4,350,854) - (6158)^2} = \frac{80,496}{1,236,722} \approx 0.065088$$

$$b = \bar{y} - m\bar{x} = \frac{\sum y}{n} - m\frac{\sum x}{n} \approx \frac{405}{9} - (0.065088)\left(\frac{6158}{9}\right) \approx 0.4653$$

$$\hat{y} = mx + b = 0.065x + 0.465$$

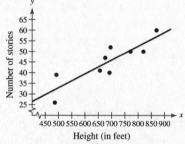

(a) $\hat{y} = 0.065(800) + 0.465 \approx 52$ stories

(b) $\hat{y} = 0.065(750) + 0.465 \approx 49$ stories

(c) It is not meaningful to predict the value of $y$ for $x = 400$ because $x = 400$ is outside the range of the original data.

(d) $\hat{y} = 0.065(625) + 0.465 \approx 41$ stories

**19.**

| $x$ | $y$ | $xy$ | $x^2$ |
|-----|-----|------|-------|
| 0 | 40 | 0 | 0 |
| 2 | 51 | 102 | 4 |
| 4 | 64 | 256 | 16 |
| 5 | 69 | 345 | 25 |
| 5 | 73 | 365 | 25 |
| 5 | 75 | 375 | 25 |
| 6 | 93 | 558 | 36 |
| 7 | 90 | 630 | 49 |
| 8 | 95 | 760 | 64 |
| $\sum x = 42$ | $\sum y = 650$ | $\sum xy = 3391$ | $\sum x^2 = 244$ |

$$m = \frac{n\sum xy - (\sum x)(\sum y)}{n\sum x^2 - (\sum x)^2} = \frac{(9)(3391) - (42)(650)}{(9)(244) - (42)^2} = \frac{3219}{432} \approx 7.451389$$

$$b = \bar{y} - m\bar{x} \approx \left(\frac{650}{9}\right) - (7.451389)\left(\frac{42}{9}\right) \approx 37.4491$$

$$\hat{y} = 7.451x + 37.449$$

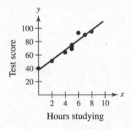

(a) $\hat{y} = 7.451(3) + 37.449 \approx 60$

(b) $\hat{y} = 7.451(6.5) + 37.449 \approx 86$

(c) It is not meaningful to predict the value of $y$ for $x = 13$ because $x = 13$ is outside the range of the original data.

(d) $\hat{y} = 7.451(4.5) + 37.449 \approx 71$

**21.**

| $x$ | $y$ | $xy$ | $x^2$ |
|---|---|---|---|
| 60 | 403 | 24,180 | 3600 |
| 75 | 363 | 27,225 | 5625 |
| 62 | 381 | 23,622 | 3844 |
| 68 | 367 | 24,956 | 4624 |
| 84 | 341 | 28,644 | 7056 |
| 97 | 317 | 30,749 | 9409 |
| 66 | 401 | 26,466 | 4356 |
| 65 | 384 | 24,960 | 4225 |
| 86 | 342 | 29,412 | 7396 |
| 78 | 377 | 29,406 | 6084 |
| 93 | 329 | 30,597 | 8649 |
| 75 | 377 | 28,275 | 5625 |
| 88 | 349 | 30,712 | 7744 |
| $\sum x = 997$ | $\sum y = 4731$ | $\sum xy = 359,204$ | $\sum x^2 = 78,237$ |

$$m = \frac{x\sum xy - (\sum x)(\sum y)}{n\sum x^2 - (\sum x)^2}$$

$$= \frac{(13)(359,204) - (997)(4731)}{(13)(78,237) - (997)^2}$$

$$= \frac{-47,155}{23,072} \approx -2.0438$$

$$b = \bar{y} - m\bar{x} \approx \left(\frac{4731}{13}\right) - (-2.04382)\left(\frac{997}{13}\right) \approx 520.6683$$

$$\hat{y} = -2.044x + 520.668$$

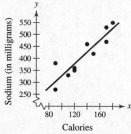

QT interval (in milliseconds) vs Heart rate (in beats per minute)

(a) It is not meaningful to predict the value of $y$ for $x = 120$ because $x = 120$ is outside the range of the original data.

(b) $\hat{y} = -2.044(67) + 520.668 = 384$ milliseconds

(c) $\hat{y} = -2.044(90) + 520.668 = 337$ milliseconds

(d) $\hat{y} = -2.044(83) + 520.668 = 351$ milliseconds

**23.**

| $x$ | $y$ | $xy$ | $x^2$ |
|---|---|---|---|
| 150 | 420 | 63,000 | 22,500 |
| 170 | 470 | 79,900 | 28,900 |
| 120 | 350 | 42,000 | 14,400 |
| 120 | 360 | 43,200 | 14,400 |
| 90 | 270 | 24,300 | 8,100 |
| 180 | 550 | 99,000 | 32,400 |
| 170 | 530 | 90,100 | 28,900 |
| 140 | 460 | 64,400 | 19,600 |
| 90 | 380 | 34,200 | 8,100 |
| 110 | 330 | 36,300 | 12,100 |
| $\sum x = 1340$ | $\sum y = 4120$ | $\sum xy = 576,400$ | $\sum x^2 = 189,400$ |

$$m = \frac{x\sum xy - (\sum x)(\sum y)}{n\sum x^2 - (\sum x)^2} = \frac{(10)(576,400) - (1340)(4120)}{(10)(189,400) - (1340)^2} = \frac{243,200}{98,400} \approx 2.471545$$

$$b = \bar{y} - m\bar{x} \approx \left(\frac{4120}{10}\right) - (2.471545)\left(\frac{1340}{10}\right) \approx 80.81301$$

$$\hat{y} = 2.472x + 80.813$$

Sodium (in milligrams) vs Calories

(a) $\hat{y} = 2.472(170) + 80.813 = 501.053$ milligrams

(b) $\hat{y} = 2.472(100) + 80.813 = 328.013$ milligrams

(c) $\hat{y} = 2.472(140) + 80.813 = 426.893$ milligrams

(d) It is not meaningful to predict the value of $y$ for $x = 210$ because $x = 210$ is outside the range of the original data.

**25.**

| $x$ | $y$ | $xy$ | $x^2$ |
|-----|-----|------|-------|
| 8.5 | 66.0 | 561.00 | 72.25 |
| 9.0 | 68.5 | 616.50 | 81.00 |
| 9.0 | 67.5 | 607.50 | 81.00 |
| 9.5 | 70.0 | 665.00 | 90.25 |
| 10.0 | 70.0 | 700.00 | 100.00 |
| 10.0 | 72.0 | 720.00 | 100.00 |
| 10.5 | 71.5 | 750.75 | 110.25 |
| 10.5 | 69.5 | 729.75 | 110.25 |
| 11.0 | 71.5 | 786.50 | 121.00 |
| 11.0 | 72.0 | 792.00 | 121.00 |
| 11.0 | 73.0 | 803.00 | 121.00 |
| 12.0 | 73.5 | 882.00 | 144.00 |
| 12.0 | 74.0 | 888.00 | 144.00 |
| 12.5 | 74.0 | 925.00 | 156.25 |
| $\sum x = 146.5$ | $\sum y = 993$ | $\sum xy = 10{,}427$ | $\sum x^2 = 1552.25$ |

$$m = \frac{x\sum xy - (\sum x)(\sum y)}{n\sum x^2 - (\sum x)^2} = \frac{(14)(10{,}427) - (146.5)(993)}{(14)(1552.25) - (146.5)^2} = \frac{503.5}{269.25} \approx 1.870009$$

$$b = \overline{y} - m\overline{x} \approx \left(\frac{993.0}{14}\right) - (1.870)\left(\frac{146.5}{14}\right) \approx 51.3603$$

$$\hat{y} = 1.870x + 51.360$$

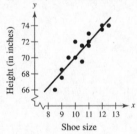

(a) $\hat{y} = 1.870(11.5) + 51.360 = 72.865$ inches

(b) $\hat{y} = 1.870(8.0) + 51.360 = 66.32$ inches

(c) It is not meaningful to predict the value of $y$ for $x = 15.5$ because $x = 15.5$ is outside the range of the original data.

(d) $\hat{y} = 1.870(10.0) + 51.360 = 70.06$ inches

**27.** Strong positive linear correlation; As the years of experience of registered nurses increase. their salaries tend to increase.

**29.** No, it is not meaningful to predict a salary for a registered nurse with 28 years of experience because $x = 28$ is outside the range of the original data.

**31. (a)**

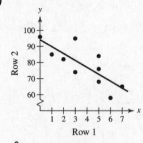

**(b)**

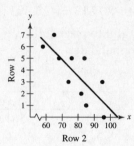

$$\hat{y} = -4.297x + 94.200 \qquad \hat{y} = -0.141x + 14.763$$

(c) The slope of the line keeps the same sign, but the values of $m$ and $b$ change.

**33. (a)** $m \approx 0.139$

$b \approx 21.024$

$\hat{y} = 0.139x + 21.024$

**(b)**

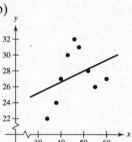

(c)

| $x$ | $y$ | $\hat{y} = 0.139x + 21.024$ | $y - \hat{y}$ |
|-----|-----|------------------------------|----------------|
| 38 | 24 | 26.306 | −2.306 |
| 34 | 22 | 22.750 | −3.750 |
| 40 | 27 | 26.584 | 0.416 |
| 46 | 32 | 27.418 | 4.582 |
| 43 | 30 | 27.001 | 2.999 |
| 48 | 31 | 27.696 | 3.304 |
| 60 | 27 | 29.364 | −2.364 |
| 55 | 26 | 28.669 | −2.669 |
| 52 | 28 | 28.252 | −0.252 |

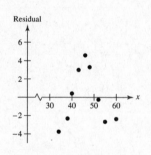

(d) The residual plot shows a pattern because the residuals do not fluctuate about 0. This implies that the regression line is not a good representation of the relationship between the variables.

**35. (a)**

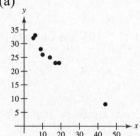

(b) The point (44, 8) may be an outlier.

(c) Excluding the point (44, 8) $\Rightarrow \hat{y} = -0.711x + 35.263$. The point (44, 8) is not influential

because using all 8 points $\Rightarrow \hat{y} = -0.607x + 34.160$.

The slopes and $y$-intercepts of the regression lines with the point included and without the point are not significantly different.

**37.** $m \approx 654.536$

$b \approx -1214.857$

$\hat{y} = 654.536x - 1214.857$

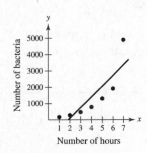

**39.** Using a technology tool $\Rightarrow y = 93.028(1.712)^x$.

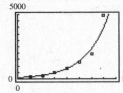

**41.** $m = -78.929$

$b = 576.179$

$\hat{y} = -78.929x + 576.179$

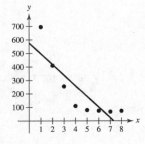

**43.** Using a technology tool $\Rightarrow y = 782.300x^{-1.251}$.

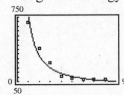

**45.** $y = a + b\ln x = 25.035 + 19.599\ln x$

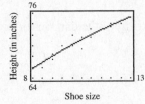

**47.** The logarithmic equation is a better model for the data. The graph of the logarithmic equation fits the data better than the regression line.

## 9.3 MEASURES OF REGRESSION AND PREDICTION INTERVALS

### 9.3 Try It Yourself Solutions

**1a.** $r \approx 0.979$      **b.** $r^2 \approx (0.979)^2 \approx 0.958$

  **c.** About 95.8% of the variation in the times is explained.
About 4.2% of the variation is unexplained.

**2a.**

| $x_i$ | $y_i$ | $\hat{y}_i$ | $y_i - \hat{y}_i$ | $\left(y_i - \hat{y}_i\right)^2$ |
|-------|-------|-------------|-------------------|----------------------------------|
| 15 | 26 | 28.386 | −2.386 | 5.692996 |
| 20 | 32 | 35.411 | −3.411 | 11.634921 |
| 20 | 38 | 35.411 | 2.589 | 6.702921 |
| 30 | 56 | 49.461 | 6.539 | 42.758521 |
| 40 | 54 | 63.511 | −9.511 | 90.459121 |
| 45 | 78 | 70.536 | 7.464 | 55.711296 |
| 50 | 80 | 77.561 | 2.439 | 5.948721 |
| 60 | 88 | 91.611 | −3.611 | 13.039321 |
|    |    |        |        | $\sum = 231.947818$ |

  **b.** $n = 8$

  **c.** $s_e = \sqrt{\dfrac{\sum\left(y_i - \hat{y}_i\right)^2}{n-2}} = \sqrt{\dfrac{231.947818}{6}} \approx 6.218$

  **d.** The standard error of estimate of the weekly sales for a specific radio ad time is about \$621.80.

**3a.** $n = 10$, d.f. $= 8$, $t_c = 2.306$, $s_e \approx 116.492$

  **b.** $\hat{y} = 166.900(4.0) + 115.725 = 783.325$

  **c.** $E = t_c s_e \sqrt{1 + \dfrac{1}{n} + \dfrac{n\left(x - \overline{x}\right)^2}{n\left(\sum x^2\right) - \left(\sum x\right)^2}}$

       $\approx (2.306)(116.492)\sqrt{1 + \dfrac{1}{10} + \dfrac{10(4.0 - 2.46)^2}{10(79.68) - (24.6)^2}}$

       $\approx 297.168$

  **d.** $\hat{y} \pm E \rightarrow 783.325 \pm 297.168 \rightarrow (486.157,\ 1080.493)$

  **e.** You can be 95% confident that when the gross domestic product is \$4 trillion, the carbon dioxide emissions will be between 486.157 and 1080.493 million metric tons.

### 9.3 EXERCISE SOLUTIONS

**1.** The total variation is the sum of the squares of the differences between the $y$-values of each ordered pair and the mean of the $y$-values of the ordered pairs or $\sum\left(y_i - \overline{y}\right)^2$.

3.  The unexplained variation is the sum of the squares of the differences between the observed
    $y$-values and the predicted $y$-values or $\sum\left(y_i - \hat{y}_i\right)^2$.

5.  Two variables that have perfect positive or perfect negative linear correlation have a correlation
    coefficient of 1 or $-1$, respectively. In either case, the coefficient of determination is 1, which
    means that 100% of the variation in the response variable is explained by the variation in the
    explanatory variable.

7.  $r^2 = (0.465)^2 \approx 0.216$; About 21.6% of the variation is explained. About 78.4% of the variation
    is unexplained.

9.  $r^2 = (-0.957)^2 \approx 0.916$; About 91.6% of the variation is explained. About 8.4% of the variation
    is unexplained.

11. (a) $r^2 = \dfrac{\sum\left(\hat{y}_i - \overline{y}\right)^2}{\sum\left(y_i - \overline{y}\right)^2} \approx 0.798$

    About 79.8% of the variation in proceeds can be explained by the relationship between the
    number of issues and proceeds, and about 20.2% of the variation is unexplained.

    (b) $s_e = \sqrt{\dfrac{\sum\left(y_i - \hat{y}_i\right)^2}{n-2}} = \sqrt{\dfrac{648,721,935}{10}} \approx 8054.328$

    The standard error of estimate of the proceeds for a specific number of issues is about
    \$8,054,328,000.

13. (a) $r^2 = \dfrac{\sum\left(\hat{y}_i - \overline{y}\right)^2}{\sum\left(y_i - \overline{y}\right)^2} \approx 0.992$

    About 99.2% of the variation in sales can be explained by the relationship between the total
    square footage and sales, and about 0.8% of the variation is unexplained.

    (b) $s_e = \sqrt{\dfrac{\sum\left(y_i - \hat{y}_i\right)^2}{n-2}} = \sqrt{\dfrac{3401.133}{9}} \approx 19.440$

    The standard error of estimate of the sales for a specific total square footage is about
    \$19,440,000,000.

15. (a) $r^2 = \dfrac{\sum\left(\hat{y}_i - \overline{y}\right)^2}{\sum\left(y_i - \overline{y}\right)^2} \approx 0.967$

    About 96.7% of the variation in average weekly wages for federal government employees can
    be explained by the relationship between average weekly wages for state government
    employees and average weekly wages for federal government employees, and about 3.3% of
    the variation is unexplained.

(b) $s_e = \sqrt{\dfrac{\sum\left(y_i - \widehat{y}_i\right)^2}{n-2}} = \sqrt{\dfrac{5060.9183}{8}} \approx 25.152$

The standard error of estimate of the average weekly wages for federal government employees for a specific average weekly wage for state government employees is about $25.15.

17. (a) $r^2 = \dfrac{\sum\left(\widehat{y}_i - \overline{y}\right)^2}{\sum\left(y_i - \overline{y}\right)^2} \approx 0.779$

About 77.9% of the variation in the amount of crude oil imported can be explained by the relationship between the amount of crude oil produced and the amount of crude oil imported, and about 22.1% of the variation is unexplained.

(b) $s_e = \sqrt{\dfrac{\sum\left(y_i - \widehat{y}_i\right)^2}{n-2}} = \sqrt{\dfrac{225{,}137.280}{5}} \approx 212.197$

The standard error of estimate of the amount of crude oil imported for a specific amount of crude oil produced is about 212,197 barrels per day.

19. (a) $r^2 = \dfrac{\sum\left(\widehat{y}_i - \overline{y}\right)^2}{\sum\left(y_i - \overline{y}\right)^2} \approx 0.899$

About 89.9% of the variation in the new-vehicle sales of General Motors can be explained by the relationship between new-vehicle sales of Ford and new-vehicle sales of General Motors , and about 10.1% of the variation is unexplained.

(b) $s_e = \sqrt{\dfrac{\sum\left(y_i - \widehat{y}_i\right)^2}{n-2}} = \sqrt{\dfrac{1{,}151{,}681.8700}{9}} \approx 357.721$

The standard error of estimate of new-vehicle sales of General Motors for a specific amount of new-vehicle sales of Ford is about 357,721 new vehicles.

21. $n = 12$, d.f. $= 10$, $t_c = 2.228$, $s_e \approx 8054.328$

$\widehat{y} = 104.965x + 14{,}093.666 = 104.965(450) + 14{,}093.666 = 61{,}327.916$

$E = t_c s_e \sqrt{1 + \dfrac{1}{n} + \dfrac{n\left(x - \overline{x}\right)^2}{n\left(\sum x^2\right) - \left(\sum x\right)^2}}$

$\approx (2.228)(8054.328)\sqrt{1 + \dfrac{1}{12} + \dfrac{12\left(450 - 2138/12\right)^2}{12(613{,}126) - (2138)^2}}$

$\approx 21{,}244.664$

$\widehat{y} \pm E \to 61{,}327.916 \pm 21{,}244.664 \to (40{,}083.251,\ 82{,}572.581)$

You can be 95% confident that the proceeds will be between $40,083,251,000 and $82,572,581,000 when the number of initial offerings is 450.

**23.** $n = 11$, d.f. $= 9$, $t_c = 1.833$, $s_e \approx 19.440$

$\hat{y} = 445.257x - 1480.117 = 445.257(5.75) - 1480.117 \approx 1080.111$

$$E = t_c s_e \sqrt{1 + \frac{1}{n} + \frac{n(x - \bar{x})^2}{n(\sum x^2) - (\sum x)^2}}$$

$$\approx (1.833)(19.440)\sqrt{1 + \frac{1}{11} + \frac{11(5.75 - 65.6/11)^2}{11(393.38) - (65.6)^2}}$$

$$\approx 37.575$$

$\hat{y} \pm E \to 1080.111 \pm 37.575 \to (1042.535, 1117.687)$

You can be 90% confident that the shopping center sales will be between \$1,042,535,000,000 and \$1,117,687,000,000 when the total square footage is 5.75 billion.

**25.** $n = 10$, d.f. $= 8$, $t_c = 3.355$, $s_e \approx 25.152$

$\hat{y} = 1.632x - 200.284 = 1.632(800) - 200.284 = 1105.316$

$$E = t_c s_e \sqrt{1 + \frac{1}{n} + \frac{n(x - \bar{x})^2}{n(\sum x^2) - (\sum x)^2}}$$

$$\approx (3.355)(25.152)\sqrt{1 + \frac{1}{10} + \frac{10(800 - 8622/10)^2}{10(7,488,884) - (8622)^2}}$$

$$\approx 91.290$$

$\hat{y} \pm E \to 1105.316 \pm 91.290 \to (1014.026, 1196.606)$

You can be 99% confident that the average weekly wages of federal government employees will be between \$1014.03 and \$1196.61 when the average weekly wage of state government employees is \$800.

**27.** $n = 7$, d.f. $= 5$, $t_c = 2.571$, $s_e \approx 212.197$

$\hat{y} = -1.167x + 16,118.76 = -1.167(5500) + 16,118.763 \approx 9700.263$

$$E = t_c s_e \sqrt{1 + \frac{1}{n} + \frac{n(x - \bar{x})^2}{n(\sum x^2) - (\sum x)^2}}$$

$$\approx (2.571)(212.197)\sqrt{1 + \frac{1}{7} + \frac{7(5500 - 37,976/7)^2}{7(206,607,544) - (37,976)^2}}$$

$$\approx 585.677$$

$\hat{y} \pm E \to 9700.263 \pm 585.677 \to (9114.586, 10,285.940)$

You can be 95% confident that the amount of crude oil imported will be between 9,114,586 and 10,285,940 barrels per day when the amount of crude oil produced is 5,500,000 barrels per day.

**29.** $n = 11$, d.f. $= 9$, $t_c = 2.262$, $s_e \approx 357.721$

$$\hat{y} = 1.200x + 433.900 = 1.200(2628) + 433.900 = 3587.500$$

$$E = t_c s_e \sqrt{1 + \frac{1}{n} + \frac{n(x - \bar{x})^2}{n\left(\sum x^2\right) - \left(\sum x\right)^2}}$$

$$\approx (2.262)(357.721)\sqrt{1 + \frac{1}{11} + \frac{11(2628 - 32,309/11)^2}{11(102,015,303) - (32,309)^2}}$$

$$\approx 850.331$$

$$\hat{y} \pm E \rightarrow 3587.500 \pm 850.331 \rightarrow (2737.169,\ 4437.831)$$

You can be 95% confident that the new-vehicle sales of General Motors will be between 2,737,169 and 4,437,831 when the new-vehicle sales of Ford is 2,628,000 new vehicles.

**31.**

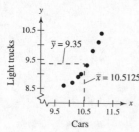

**33.** $r^2 \approx 0.934$; About 93.4% of the variation in the median ages of light trucks can be explained by the relationship between the median ages of light trucks and the median ages of cars, and about 6.6% of the variation is unexplained.

**35.** critical value $= \pm 3.707$

$m \approx -0.205$

$s_e \approx 0.554$

$$t = \frac{m}{s_e}\sqrt{\sum x^2 - \frac{\left(\sum x\right)^2}{n}} \approx \frac{-0.205}{0.554}\sqrt{838.55 - \frac{(79.5)^2}{8}} \approx -2.578$$

Because $-3.707 < t < 3.707$, fail to reject $H_0 : M = 0$. There is not enough evidence at the 1% level of significance to support the claim that there is a linear relationship between weight and number of hours slept.

**37.** $E = t_c s_e \sqrt{\dfrac{1}{n} + \dfrac{\bar{x}^2}{\sum x^2 - \left[\left(\sum x\right)^2 / n\right]}} \approx (2.306)(116.492)\sqrt{\dfrac{1}{10} + \dfrac{\left(\dfrac{24.6}{10}\right)^2}{(79.68) - \dfrac{(24.6)^2}{10}}} \approx 173.216$

$$b \pm E \rightarrow 115.725 \pm 173.216 \rightarrow (-57.491,\ 288.941)$$

$$E = \frac{t_c s_e}{\sqrt{\sum x^2 - \dfrac{(\sum x)^2}{n}}} \approx \frac{2.306(116.492)}{\sqrt{79.68 - \dfrac{(24.6)^2}{10}}} \approx 61.364$$

$$m \pm E \rightarrow 166.900 \pm 61.364 \rightarrow (105.536, \ 228.264)$$

## 9.4 MULTIPLE REGRESSION

## 9.4 Try It Yourself Solutions

**1a.** Enter the data.　　　　**b.** $\hat{y} = 46.385 + 0.540x_1 - 4.897x_2$

**2ab.** (1) $\hat{y} = 46.385 + 0.540(89) - 4.897(1)$

(2) $\hat{y} = 46.385 + 0.540(78) - 4.897(3)$

(3) $\hat{y} = 46.385 + 0.540(83) - 4.897(2)$

**c.** (1) $\hat{y} = 89.548$　　　(2) $\hat{y} = 73.814$　　　(3) $\hat{y} = 81.411$

**d.** (1) 90　　　(2) 74　　　(3) 81

## 9.4 EXERCISE SOLUTIONS

1. $\hat{y} = 24{,}791 + 4.508x_1 - 4.723x_2$

(a) $\hat{y} = 24{,}791 + 4.508(36{,}500) - 4.723(36{,}100) = 18{,}832.7$ pounds per acre

(b) $\hat{y} = 24{,}791 + 4.508(38{,}100) - 4.723(37{,}800) = 18{,}016.4$ pounds per acre

(c) $\hat{y} = 24{,}791 + 4.508(39{,}000) - 4.723(38{,}800) = 17{,}350.6$ pounds per acre

(d) $\hat{y} = 24{,}791 + 4.508(42{,}200) - 4.723(42{,}100) = 16{,}190.3$ pounds per acre

3. $\hat{y} = -52.2 + 0.3x_1 + 4.5x_2$

(a) $\hat{y} = -52.2 + 0.3(70) + 4.5(8.6) = 7.5$ cubic feet

(b) $\hat{y} = -52.2 + 0.3(65) + 4.5(11.0) = 16.8$ cubic feet

(c) $\hat{y} = -52.2 + 0.3(83) + 4.5(17.6) = 51.9$ cubic feet

(d) $\hat{y} = -52.2 + 0.3(87) + 4.5(19.6) = 62.1$ cubic feet

5. $\hat{y} = -2075.2 + 20.9x_1 + 35.071x_2$

(a) $s_e = 8.721$; The standard error of estimate of the predicted sales given a specific total square footage and number of shopping centers is \$8.721 billion.

(b) $r^2 = 0.998$; The multiple regression model explains about 99.8% of the variation in $y$.

**7.** $n = 8 \ k = 2, \ r^2 = 0.998$

$$r_{adj}^2 = 1 - \left[ \frac{(1 - r^2)(n - 1)}{n - k - 1} \right] \approx 0.997$$

About 9.7% of the variation in $y$ can be explained by the relationships between variables.
$r_{adj}^2 < r^2$

# CHAPTER 9 REVIEW EXERCISE SOLUTIONS

**1.** (a)

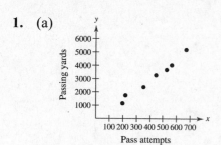

(b)

| $x$ | $y$ | $xy$ | $x^2$ | $y^2$ |
|---|---|---|---|---|
| 449 | 3265 | 1,465,985 | 201,601 | 10,660,225 |
| 565 | 4018 | 2,270,170 | 319,225 | 16,144,324 |
| 528 | 3669 | 1,937,232 | 278,784 | 13,461,561 |
| 197 | 1141 | 224,777 | 38,809 | 1,301,881 |
| 670 | 5177 | 3,468,590 | 448,900 | 26,801,329 |
| 351 | 2362 | 829,062 | 123,201 | 5,579,044 |
| 218 | 1737 | 378,666 | 47,524 | 3,017,169 |
| $\sum x = 2978$ | $\sum y = 21,369$ | $\sum xy = 10,574,482$ | $\sum x^2 = 1,458,044$ | $\sum y^2 = 76,965,533$ |

$$r = \frac{n\sum xy - (\sum x)(\sum y)}{\sqrt{n\sum x^2 - (\sum x)^2} \ \sqrt{n\sum y^2 - (\sum y)^2}}$$

$$= \frac{7(10,574,482) - (2978)(21,369)}{\sqrt{7(1,458,044) - (2978)^2} \ \sqrt{7(76,965,533) - (21,369)^2}}$$

$$= \frac{10,384,492}{\sqrt{1,337,824} \ \sqrt{82,124,570}} \approx 0.991$$

(c) Strong positive linear correlation; As the number of pass attempts increases, the number of passing yards tends to increase.

**3.** (a)

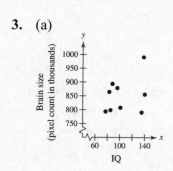

(b)

| $x$ | $y$ | $xy$ | $x^2$ | $y^2$ |
|---|---|---|---|---|
| 138 | 991 | 136,758 | 19,044 | 982081 |
| 140 | 856 | 119,840 | 1960 | 732736 |
| 96 | 879 | 84,384 | 9216 | 772641 |
| 83 | 865 | 71,795 | 6889 | 748225 |
| 101 | 808 | 81,608 | 10,201 | 652864 |
| 135 | 791 | 106,785 | 18,225 | 625681 |
| 85 | 799 | 67,915 | 7225 | 638401 |
| 77 | 794 | 61,138 | 5929 | 630436 |
| 88 | 894 | 78,672 | 7744 | 799236 |
| $\sum x = 943$ | $\sum y = 7677$ | $\sum xy = 808,895$ | $\sum x^2 = 104,073$ | $\sum y^2 = 6,582,301$ |

$$r = \frac{n\sum xy - (\sum x)(\sum y)}{\sqrt{n\sum x^2 - (\sum x)^2}\sqrt{n\sum y^2 - (\sum y)^2}}$$

$$= \frac{9(808,865) - (943)(7677)}{\sqrt{9(104,073) - (943)^2}\sqrt{9(6,582,301) - (7677)^2}}$$

$$= \frac{40,644}{\sqrt{47,408}\sqrt{304,380}} \approx 0.338$$

(c) Weak positive linear correlation; The IQ does not appear to be related to the brain size.

5. $H_0: \rho = 0;\ H_a: \rho \neq 0$

$\alpha = 0.05$, d.f. $= n - 2 = 5$

$-t_0 = -2.571,\ t_0 = 2.571$; Rejection regions: $t < -2.571$ or $t > 2.086$

$$t = \frac{r}{\sqrt{\dfrac{1 - r^2}{n - 2}}} = \frac{0.991}{\sqrt{\dfrac{1 - (0.991)^2}{7 - 2}}} \approx 16.554$$

Because $t > 2.086$, reject $H_0$. There is enough evidence at the 5% level of significance to conclude that a significant linear correlation exists between a quarterback's passing attempts and passing yards.

7. $H_0: \rho = 0;\ H_a: \rho \neq 0$

$\alpha = 0.01$, d.f. $= n - 2 = 7$

$-t_0 = -3.499,\ t_0 = 3.499$; Rejection regions: $t < -3.499$ or $t > 3.499$

$$t = \frac{r}{\sqrt{\dfrac{1 - r^2}{n - 2}}} = \frac{0.338}{\sqrt{\dfrac{1 - (0.338)^2}{9 - 2}}} \approx 0.950$$

Because $-3.499 < t < 3.499$, fail to reject $H_0$. There is not enough evidence at the 1% level of significance to conclude that a significant linear correlation between IQ and brain size.

**9.** $\hat{y} = 0.038x - 3.529$

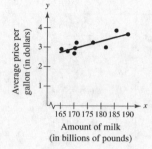

(a) It is not meaningful to predict the value of $y$ for $x = 150$ because $x = 150$ is outside the range of the original data.

(b) $\hat{y} = 0.038(175) - 3.529 = 3.121$ or \$3.12

(c) $\hat{y} = 0.038(180) - 3.529 = 3.311$ or \$3.31

(d) It is not meaningful to predict the value of $y$ for $x = 210$ because $x = 210$ is outside the range of the original data.

**11.** $\hat{y} = -0.086x + 10.450$

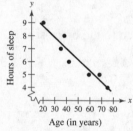

(a) It is not meaningful to predict the value of $y$ for $x = 16$ because $x = 16$ is outside the range of the original data.

(b) $\hat{y} = -0.086(25) + 10.450 = 8.3$ hours

(c) It is not meaningful to predict the value of $y$ for $x = 85$ because $x = 85$ is outside the range of the original data.

(d) $\hat{y} = -0.086(50) + 10.450 = 6.15$ hours

**13.** $r^2 = (-0.450)^2 \approx 0.203$

About 20.3% of the variation in $y$ is explained.
About 79.7% of the variation in $y$ is unexplained.

**15.** $r^2 = (0.642)^2 \approx 0.412$

About 41.2% of the variation in $y$ is explained.
About 58.8% of the variation in $y$ is unexplained.

**17.** (a) $r^2 = \dfrac{\sum\left(\hat{y}_i - \bar{y}\right)^2}{\sum\left(y_i - \bar{y}\right)^2} \approx 0.679$

About 67.9% of the variation in the fuel efficiency of the compact sports sedans cans be explained by the relationship between price and the fuel efficiency, and about 32.1% is unexplained.

(b) $s_e = \sqrt{\dfrac{\sum\left(y_i - \widehat{y}_i\right)^2}{n-2}} = \sqrt{\dfrac{9.0633}{7}} \approx 1.138$

The standard error of estimate of the fuel efficiency of compact sports sedans for a specific price of compact sports sedan is about 1.138 miles per gallon.

**19.** $\widehat{y} = 0.038(185) - 3.529 = 3.501$

$E = t_c s_e \sqrt{1 + \dfrac{1}{n} + \dfrac{n\left(x - \overline{x}\right)^2}{n\left(\sum x^2\right) - \left(\sum x\right)^2}} \approx (1.895)(0.246)\sqrt{1 + \dfrac{1}{9} + \dfrac{9(185 - 1578.8/9)^2}{9(277,555.56) - (1578.8)^2}}$

$\approx 0.524$

$\widehat{y} \pm E \rightarrow 3.501 \pm 0.524 \rightarrow (2.997,\ 4.025)$

You can be 90% confident that the price per gallon of milk will be between \$3.00 and \$4.03 when 185 billion pounds of milk is produced.

**21.** $\widehat{y} = -0.086(45) + 10.450 = 6.580$

$E = t_c s_e \sqrt{1 + \dfrac{1}{n} + \dfrac{n\left(x - \overline{x}\right)^2}{n\left(\sum x^2\right) - \left(\sum x\right)^2}} \approx (2.571)(0.622)\sqrt{1 + \dfrac{1}{7} + \dfrac{7(45 - 337/7)^2}{7(18,563) - (337)^2}}$

$\approx 1.714$

$\widehat{y} \pm E \rightarrow 6.580 \pm 1.714 \rightarrow (4.866,\ 8.294)$

You can be 95% confident that the hours slept will be between 4.866 and 8.294 hours for a person who is 45 years old.

**23.** $\widehat{y} = -0.414(39.9) + 37.147 \approx 20.628$

$E = t_c s_e \sqrt{1 + \dfrac{1}{n} + \dfrac{n\left(x - \overline{x}\right)^2}{n\left(\sum x^2\right) - \left(\sum x\right)^2}} \approx (3.499)(1.138)\sqrt{1 + \dfrac{1}{9} + \dfrac{9(39.9 - 319.7/9)^2}{9(11,468.29) - (319.7)^2}}$

$\approx 4.509$

$\widehat{y} \pm E \rightarrow 20.628 \pm 4.509 \rightarrow (16.119,\ 25.137)$

You can be 99% confident that the fuel efficiency of the compact sports sedan that costs \$39,900 will be between 16.119 and 25.137 miles per gallon.

**25.** (a) $\widehat{y} = 3.6738 + 1.2874x_1 - 7.531x_2$

(b) $s_e \approx 0.710$; The standard error of estimate of the carbon monoxide content given specific tar and nicotine contents is about 0.710 milligram.

(c) $r^2 \approx 0.943$; The multiple regression model explains about 94.3% of the variation in $y$.

**27.** (a) 21.705 miles per gallon      (b) 25.21 miles per gallon

(c) 30.1 miles per gallon          (d) 25.86 miles per gallon

## CHAPTER 9 QUIZ SOLUTIONS

1.

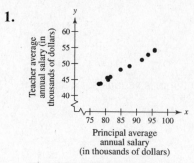

Principal average
annual salary
(in thousands of dollars)

The data appear to have a positive linear correlation. As $x$ increases, $y$ tends to increase.

2. $r \approx 0.998$; Strong positive linear correlation; As the average annual salaries of public school principals increase, the average annual salaries of public school classroom teachers tend to increase.

3. $H_0 : \rho = 0; H_a : \rho \neq 0$

$\alpha = 0.05$, d.f. $= n - 2 = 9$

$-t_0 = -2.262, t_0 = 2.262$;  Rejection regions: $t < -2.262$ or $t > 2.262$

$$t = \frac{r}{\sqrt{\dfrac{1-r^2}{n-2}}} \approx \frac{0.998}{\sqrt{\dfrac{1-(0.998)^2}{11-2}}} \approx 47.36$$

Because $t > 2.262$, reject $H_0$. There is enough evidence at the 5% level of significance to conclude that there is a significant linear correlation between the average annual salaries of public school principals and the average annual salaries of public school classroom teachers.

4.

| $x$ | $y$ | $xy$ | $x^2$ |
|---|---|---|---|
| 77.8 | 43.7 | 3399.86 | 6052.84 |
| 78.4 | 43.8 | 3433.92 | 6146.56 |
| 80.8 | 45.0 | 3636.00 | 6528.64 |
| 80.5 | 45.6 | 3670.80 | 6480.25 |
| 81.5 | 45.9 | 3740.85 | 6642.25 |
| 84.8 | 48.2 | 4087.36 | 7191.04 |
| 87.7 | 49.3 | 4323.61 | 7691.29 |
| 91.6 | 51.3 | 4699.08 | 8390.56 |
| 93.6 | 52.9 | 4951.44 | 8760.96 |
| 95.7 | 54.4 | 5206.08 | 9158.49 |
| 95.7 | 54.2 | 5186.94 | 9158.49 |
| $\sum x = 948.1$ | $\sum y = 534.3$ | $\sum xy = 46{,}335.94$ | $\sum x^2 = 82{,}201.37$ |

$$m = \frac{n\sum xy - (\sum x)(\sum y)}{n\sum x^2 - (\sum x)^2} = \frac{11(46{,}335.94) - (948.1)(534.3)}{11(82{,}201.37) - (948.1)^2} = \frac{3125.51}{5321.46} \approx 0.5873407$$

$$b = \bar{y} - m\bar{x} = \frac{\sum y}{n} - m\left(\frac{\sum x}{n}\right) \approx \frac{534.3}{11} - (0.5873407)\left(\frac{948.1}{11}\right) \approx -2.051$$

$$\hat{y} = mx + b = 0.587x - 2.051$$

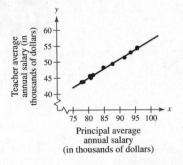

5. $\hat{y} = 0.587(90.5) - 2.051 = 51.0725 \Rightarrow \$51,072.50$

6. $r^2 \approx (0.9975)^2 \approx 0.995$

   About 99.5% of the variation in the average annual salaries of public school classroom teachers can be explained by the relationship between the average annual salaries of public school principals and the average annual salaries of public school classroom teachers, and about 0.6% of the variation is unexplained.

7. $s_e \approx 0.307$; The standard error of the estimate of the average annual salary of public school classroom teachers for a specific average annual salary of public school principals is about \$307.

8. $\hat{y} = 0.587(85.75) - 2.051 = 48.284$

$$E = t_c s_e \sqrt{1 + \frac{1}{n} + \frac{n(x - \bar{x})^2}{n(\sum x^2) - (\sum x)^2}} \approx (2.262)(0.307)\sqrt{1 + \frac{1}{11} + \frac{11(85.75 - 948.1/11)^2}{11(82,201.37) - (948.1)^2}}$$

$$\approx 0.725$$

$$\hat{y} \pm E \rightarrow 48.284 \pm 0.725 \rightarrow (47.559,\ 49.009)$$

   You can be 95% confident that the average annual salary of public school classroom teachers will be between \$47,559 and \$49,009 when the average annual salary of public school principals is \$85,750.

9. (a) $\hat{y} = -86 + 7.46(27.6) - 1.61(15.3) = \$95.26$

   (b) $\hat{y} = -86 + 7.46(24.1) - 1.61(14.6) = \$70.28$

   (c) $\hat{y} = -86 + 7.46(23.5) - 1.61(13.4) = \$67.74$

   (d) $\hat{y} = -86 + 7.46(22.8) - 1.61(15.3) = \$59.46$

## 10.1 GOODNESS OF FIT TEST

## 10.1 Try It Yourself Solutions

**1.**

| Tax Preparation Method | % of people | Expected frequency |
|---|---|---|
| Accountant | 24% | $500(0.24) = 120$ |
| By hand | 20% | $500(0.20) = 100$ |
| Computer software | 35% | $500(0.35) = 175$ |
| Friend/family | 6% | $500(0.06) = 30$ |
| Tax preparation service | 15% | $500(0.15) = 75$ |

**2a.** The expected frequencies are 64, 80, 32, 56, 60, 48, 40, and 20, all of which are at least 5.

**b.** Claimed distribution:

| Ages | Distribution |
|---|---|
| 0–9 | 16% |
| 10–19 | 20% |
| 20–29 | 8% |
| 30–39 | 14% |
| 40–49 | 15% |
| 50–59 | 12% |
| 60–69 | 10% |
| 70+ | 5% |

$H_0$ : The distribution of ages is as shown in table above.

$H_a$ : The distribution of ages differs from the claimed distribution. (claim)

**c.** $\alpha = 0.05$  **d.** d.f. $= n - 1 = 7$

**e.** $\chi_0^2 = 14.067$ ; Rejection region: $\chi^2 > 14.067$

**f.**

| Ages | Distribution | Observed | Expected | $\dfrac{(O - E)^2}{E}$ |
|---|---|---|---|---|
| 0–9 | 16% | 76 | 64 | 2.250 |
| 10–19 | 20% | 84 | 80 | 0.200 |
| 20–29 | 8% | 30 | 32 | 0.125 |
| 30–39 | 14% | 60 | 56 | 0.286 |
| 40–49 | 15% | 54 | 60 | 0.600 |
| 50–59 | 12% | 40 | 48 | 1.333 |
| 60–69 | 10% | 42 | 40 | 0.100 |
| 70+ | 5% | 14 | 20 | 1.800 |

$\chi^2 \approx 6.694$

$\alpha = 0.05$

$\chi^2 \approx 6.694$

**g.** Because $\chi^2 < 14.067$, fail to reject $H_0$.

**h.** There is not enough evidence at the 5% level of significance to support the sociologist's claim that the distribution of ages differs from the age distribution 10 years ago.

**3a.** The expected frequency for each category is 30 which is at least 5.

**b.** Claimed distribution:

| Color | Distribution |
|-------|--------------|
| Brown | 16.$\overline{6}$% |
| Yellow | 16.$\overline{6}$% |
| Red | 16.$\overline{6}$% |
| Blue | 16.$\overline{6}$% |
| Orange | 16.$\overline{6}$% |
| Green | 16.$\overline{6}$% |

$H_0$ : The distribution of colors is uniform, as shown in the table above. (claim)

$H_a$ : The distribution of colors is not uniform.

**c.** $\alpha = 0.05$

**d.** d.f. $= n - 1 = 5$

**e.** $\chi_0^2 = 11.071$; Rejection region: $\chi^2 > 11.071$

**f.**

| Color | Distribution | Observed | Expected | $\dfrac{(O - E)^2}{E}$ |
|-------|--------------|----------|----------|------------------------|
| Brown | 16.6% | 22 | 30 | 2.1$\overline{33}$ |
| Yellow | 16.6% | 27 | 30 | 0.300 |
| Red | 16.6% | 22 | 30 | 2.1$\overline{33}$ |
| Blue | 16.6% | 41 | 30 | 4.0$\overline{33}$ |
| Orange | 16.6% | 41 | 30 | 4.0$\overline{33}$ |
| Green | 16.6% | 27 | 30 | 0.300 |
| | | | | 12.933 |

$\chi^2 \approx 12.933$

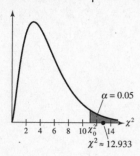

$\chi^2 \approx 12.933$

**g.** Because $\chi^2 > 11.071$, reject $H_0$.

**h.** There is enough evidence at the 5% level of significance to dispute the claim that the distribution of different colored candies in bags of peanut M&M's is uniform.

## 10.1 EXERCISE SOLUTIONS

1.  A multinomial experiment is a probability experiment consisting of a fixed number of independent trials in which there are more than two possible outcomes for each trial. The probability of each outcome is fixed, and each outcome is classified into categories.

3.  $E_i = np_i = (150)(0.3) = 45$

5.  $E_i = np_i = (230)(0.25) = 57.5$

7.  (a) Claimed Distribution:

| Age | Distribution |
|-----|--------------|
| 2–17 | 22% |
| 18–24 | 21% |
| 25–39 | 24% |
| 40–49 | 14% |
| 50+ | 19% |

$H_0$ : The distribution of the ages of moviegoers is 22% ages 2–17, 21% ages 18–24, 24% ages 25–39, 14% ages 40–49, and 19% ages 50+ (claim)

$H_a$ : The distribution of the ages differs from the claimed or expected distribution.

(b) $\chi_0^2 = 7.779$, Rejection region: $\chi^2 > 7.779$

(c)

| Age | Distribution | Observed | Expected | $\dfrac{(O-E)^2}{E}$ |
|-----|--------------|----------|----------|----------------------|
| 2–17 | 22% | 240 | 220 | 1.8182 |
| 18–24 | 21% | 214 | 210 | 0.0762 |
| 25–39 | 24% | 183 | 240 | 13.5375 |
| 40–49 | 14% | 156 | 140 | 1.8286 |
| 50+ | 19% | 207 | 190 | 1.5211 |
| | | | | 18.7815 |

$\chi^2 \approx 18.7815$

(d) Because $\chi^2 > 7.779$, reject $H_0$.

(e) There is enough evidence at the 10% level of significance to reject the claim that the distribution of the ages of moviegoers and the expected distribution are the same.

9.  (a) Claimed distribution:

| Day | Distribution |
|-----------|:---:|
| Sunday | 7% |
| Monday | 4% |
| Tuesday | 6% |
| Wednesday | 13% |
| Thursday | 10% |
| Friday | 36% |
| Saturday | 24% |

$H_0$ : The distribution of the days people order food for delivery is 7% Sunday, 4% Monday, 6% Tuesday, 13% Wednesday, 10% Thursday, 36% Friday, and 24% Saturday.

$H_a$ : The distribution of days differs from the claimed or expected distribution.

(b) $\chi_0^2 = 16.812$ , Rejection region: $\chi^2 > 16.812$

(c)

| | Distribution | Observed | Expected | $\dfrac{(O-E)^2}{E}$ |
|-----------|:---:|:---:|:---:|:---:|
| Sunday | 7% | 43 | 35 | 1.8286 |
| Monday | 4% | 16 | 20 | 0.8000 |
| Tuesday | 6% | 25 | 30 | 0.8333 |
| Wednesday | 13% | 49 | 65 | 3.9385 |
| Thursday | 10% | 46 | 50 | 0.3200 |
| Friday | 36% | 168 | 180 | 0.8000 |
| Saturday | 24% | 153 | 120 | 9.0750 |
| | | | | 17.5954 |

$\chi^2 \approx 17.5954$

(d) Reject $H_0$ .

(e) There is enough evidence at the 1% level of significance to conclude that the distribution of delivery days has changed.

11. (a) Claimed distribution:

| Season | Distribution |
|--------|:---:|
| Spring | 25% |
| Summer | 25% |
| Fall | 25% |
| Winter | 25% |

$H_0$ : The number of homicide crimes in California by season is uniform. (claim)

$H_a$ : The number of homicide crimes in California by season is not uniform.

(b) $\chi_0^2 = 7.815$ , Rejection region: $\chi^2 > 7.815$

(c)

| Season | Distribution | Observed | Expected | $\dfrac{(O-E)^2}{E}$ |
|--------|--------------|----------|----------|--------|
| Spring | 25% | 309 | 300 | 0.2700 |
| Summer | 25% | 312 | 300 | 0.4800 |
| Fall | 25% | 290 | 300 | 0.3333 |
| Winter | 25% | 289 | 300 | 0.4033 |
| | | | | 1.4866 |

$\chi^2 \approx 1.487$

(d) Because $\chi^2 < 7.815$, fail to reject $H_0$.

(e) There is not enough evidence at the 5% level of significance to reject the claim that the distribution of the number of homicide crimes in California by season is uniform.

13. (a) Claimed distribution:

| Month | Distribution |
|-------|--------------|
| Strongly agree | 55% |
| Somewhat agree | 30% |
| Neither agree nor disagree | 5% |
| Somewhat disagree | 6% |
| Strongly disagree | 4% |

$H_0$: The distribution of the opinions of U.S. parents on whether a college education is worth the expense is 55% strongly agree, 30% somewhat agree, 5% neither agree nor disagree, 6% somewhat disagree, 4% strongly disagree.

$H_a$: The distribution of opinions differs from the expected distribution. (claim)

(b) $\chi_0^2 = 9.488$, Rejection region: $\chi^2 > 9.488$

(c)

| Month | Distribution | Observed | Expected | $\dfrac{(O-E)^2}{E}$ |
|-------|--------------|----------|----------|--------|
| Strongly agree | 55% | 86 | 110 | 5.2364 |
| Somewhat agree | 30% | 62 | 60 | 0.0667 |
| Neither agree nor disagree | 5% | 34 | 10 | 57.6000 |
| Somewhat disagree | 6% | 14 | 12 | 0.3333 |
| Strongly disagree | 4% | 4 | 8 | 2.0000 |
| | | | | 65.2364 |

$\chi^2 \approx 65.236$

(d) Because $\chi^2 > 9.488$, reject $H_0$.

(e) There is enough evidence at the 5% level of significance to conclude that the distribution of opinions of U.S. parents on whether a college education is worth the expense differs from the claimed or expected distribution.

15. (a) Claimed distribution:

| Response | Distribution |
|----------|--------------|
| Larger | $33.\overline{3}\%$ |
| Same size | $33.\overline{3}\%$ |
| Smaller | $33.\overline{3}\%$ |

$H_0$ : The distribution of prospective home buyers by the size they want their next house to be is uniform.

$H_a$ : The distribution of prospective home buyers by the size they want their next house to be is not uniform. (claim)

(b) $\chi_0^2 = 5.991$, Rejection region: $\chi^2 > 5.991$

(c)

| Response | Distribution | Observed | Expected | $\dfrac{(O-E)^2}{E}$ |
|---|---|---|---|---|
| Larger | $33.\overline{3}\%$ | 285 | $266.\overline{66}$ | 1.2604 |
| Same size | $33.\overline{3}\%$ | 224 | $266.\overline{66}$ | 6.8267 |
| Smaller | $33.\overline{3}\%$ | 291 | $266.\overline{66}$ | 2.2204 |
| | | | | 10.3075 |

$\chi^2 \approx 10.308$

(d) Because $\chi^2 > 5.991$, reject $H_0$.

(e) There is enough evidence at the 5% level of significance to conclude that the distribution of prospective home buyers by the size they want their next house to be is not uniform.

**17.** (a) Frequency distribution: $\mu = 69.435$; $\sigma \approx 8.337$

| Lower Boundary | Upper Boundary | Lower *z*-score | Upper *z*-score | Area |
|---|---|---|---|---|
| 49.5 | 58.5 | −2.39 | −1.31 | 0.0867 |
| 58.5 | 67.5 | −1.31 | −0.23 | 0.3139 |
| 67.5 | 76.5 | −0.23 | 0.85 | 0.3933 |
| 76.5 | 85.5 | 0.85 | 1.93 | 0.1709 |
| 85.5 | 94.5 | 1.93 | 3.01 | 0.0255 |

| Class Boundaries | Distribution | Frequency | Expected | $\dfrac{(O-E)^2}{E}$ |
|---|---|---|---|---|
| 49.5–58.5 | 8.67% | 19 | 17 | 0.2353 |
| 58.5–67.5 | 31.39% | 61 | 63 | 0.0635 |
| 67.5–76.5 | 39.33% | 82 | 79 | 0.1139 |
| 76.5–85.5 | 17.09% | 34 | 34 | 0 |
| 85.5–94.5 | 2.55% | 4 | 5 | 0.2000 |
| | | 200 | | 0.6127 |

$H_0$ : Test scores have a normal distribution. (claim)

$H_a$ : Test scores do not have a normal distribution.

(b) $\chi_0^2 = 13.277$; Rejection region . $\chi^2 > 13.277$

(c) $\chi^2 = 0.613$

(d) Because $\chi^2 < 13.277$, fail to reject $H_0$.

(e) There is not enough evidence at the 1% level of significance to reject the claim that the test scores are normally distributed.

## 10.2 INDEPENDENCE

### 10.2 Try It Yourself Solutions

**1a.**

|  | Hotel | Leg Room | Rental Size | Other | Total |
|---|---|---|---|---|---|
| Business | 36 | 108 | 14 | 22 | 180 |
| Leisure | 38 | 54 | 14 | 14 | 120 |
| Total | 74 | 162 | 28 | 36 | 300 |

**b.** $n = 300$

**c.**

|  | Hotel | Leg Room | Rental Size | Other |
|---|---|---|---|---|
| Business | 44.4 | 97.2 | 16.8 | 21.6 |
| Leisure | 29.6 | 64.8 | 11.2 | 14.4 |

**2a.** The claim is that "the travel concerns depend on the purpose of travel."

$H_0$ : Travel concern is independent of travel purpose.

$H_a$ : Travel concern is dependent on travel purpose. (claim)

**b.** $\alpha = 0.01$          **c.** $(r-1)(c-1) = 3$

**d.** $\chi_0^2 = 11.345$; Rejection region: $\chi^2 > 11.345$

**e.**

| $O$ | $E$ | $O - E$ | $(O - E)^2$ | $\dfrac{(O - E)^2}{E}$ |
|---|---|---|---|---|
| 36 | 44.4 | −8.4 | 70.56 | 1.5892 |
| 108 | 97.2 | 10.8 | 116.64 | 1.2000 |
| 14 | 16.8 | −2.8 | 7.84 | 0.4667 |
| 22 | 21.6 | 0.4 | 0.16 | 0.0074 |
| 38 | 29.6 | 8.4 | 70.56 | 2.3838 |
| 54 | 64.8 | −10.8 | 116.64 | 1.8000 |
| 14 | 11.2 | 2.8 | 7.84 | 0.7000 |
| 14 | 14.4 | −0.4 | 0.16 | 0.0111 |
|  |  |  |  | 8.1582 |

$\chi^2 \approx 8.158$

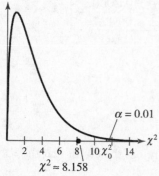

$\chi^2 \approx 8.158$

**f.** Because $\chi^2 < 11.345$, fail to reject $H_0$.

**g.** There is not enough evidence at the 1% level of significance for the consultant to conclude that travel concern is dependent on travel purpose.

**3a.** $H_0$ : Whether or not a tax cut would influence an adult to purchase a hybrid vehicle is independent of age.

   $H_a$ : Whether or not a tax cut would influence an adult to purchase a hybrid vehicle is dependent on age. (claim)

**b.** Enter the data.

**c.** $\chi_0^2 = 9.210$; Rejection region: $\chi^2 > 9.210$

**d.** $\chi^2 \approx 15.306$

**e.** Because $\chi^2 > 9.210$, reject $H_0$.

**f.** There is enough evidence at the 1% level of significance to conclude that whether or not a tax cut would influence an adult to purchase a hybrid vehicle is dependent on age.

# 10.2 EXERCISE SOLUTIONS

**1.** Find the sum of the row and the sum of the column in which the cell is located. Find the product of these sums. Divide the product by the sample size.

**3.** Answer will vary. *Sample answer:* For both the chi-square test for independence and the chi-square goodness-of-fit test, you are testing a claim about data that are in categories. However, the chi-square goodness-of-fit test has only one data value per category, while the chi-square test for independence has multiple data values per category.

Both tests compare observed and expected frequencies. However, the chi-square goodness-of-fit test simply compares the distributions, whereas the chi-square test for independence compares them and then draws a conclusion about the dependence or independence of the variables.

**5.** False. If the two variables of a chi-square test for independence are dependent, then you can expect a large difference between the observed frequencies and the expected frequencies.

**7.** (a)

| Result | Athlete has Stretched | Not stretched | Total |
|--------|-----------|---------------|-------|
| Inquiry | 18 | 22 | 40 |
| No Inquiry | 211 | 189 | 400 |
| | 229 | 211 | 440 |

(b)

| Result | Athlete has Stretched | Not stretched |
|--------|-----------|---------------|
| Inquiry | 20.82 | 19.18 |
| No Inquiry | 208.18 | 191.82 |

**9. (a)**

| Bank employee | Preference | | | Total |
|---|---|---|---|---|
| | **New procedure** | **Old Procedure** | **No preference** | |
| Teller | 92 | 351 | 50 | 493 |
| Customer service | 76 | 42 | 8 | 126 |
| Total | 168 | 393 | 58 | 619 |

**(b)**

| Bank employee | Preference | | |
|---|---|---|---|
| | **New procedure** | **Old Procedure** | **No preference** |
| Teller | 133.80 | 313.00 | 46.19 |
| Customer service | 34.20 | 80.00 | 11.81 |

**11. (a)**

| Gender | Type of car | | | | Total |
|---|---|---|---|---|---|
| | **Compact** | **Full-size** | **SUV** | **Truck/Van** | |
| Male | 28 | 39 | 21 | 22 | 110 |
| Female | 24 | 32 | 20 | 14 | 90 |
| | 52 | 71 | 41 | 36 | 200 |

**(b)**

| Gender | Type of car | | | |
|---|---|---|---|---|
| | **Compact** | **Full-size** | **SUV** | **Truck/Van** |
| Male | 28.6 | 39.05 | 22.55 | 19.8 |
| Female | 23.4 | 31.95 | 18.45 | 16.2 |

**13. (a)** The claim is "achieving a basic skill level is related to the location of the school."

$H_0$ : Skill level in a subject is independent of location. (claim)

$H_a$ : Skill level in a subject is dependent on location.

**(b)** d.f. $= (r-1)(c-1) = 2$

$\chi_0^2 = 9.210$ ; Rejection region: $\chi^2 > 9.210$

**(c)**

| $O$ | $E$ | $O-E$ | $(O-E)^2$ | $\dfrac{(O-E)^2}{E}$ |
|---|---|---|---|---|
| 43 | 41.129 | 1.871 | 3.500641 | 0.0851 |
| 42 | 41.905 | 0.095 | 0.009025 | 0.0002 |
| 38 | 39.965 | −1.965 | 3.861225 | 0.0966 |
| 63 | 64.871 | −1.871 | 3.500641 | 0.0540 |
| 66 | 66.095 | −0.095 | 0.009025 | 0.0001 |
| 65 | 63.035 | 1.965 | 3.861225 | 0.0613 |
| | | | | 0.2973 |

$\chi^2 \approx 0.297$

**(d)** Because $\chi^2 < 9.210$, fail to reject $H_0$.

(e) There is not enough evidence at the 1% level of significance to reject the claim that skill level in a subject is independent of location.

**15.** (a) The claim is "the number of times former smokers tried to quit before they were habit-free is related to gender."

$H_0$ : The number of times former smokers tried to quit is independent of gender.

$H_a$ : The number of times former smokers tried to quit is dependent of gender. (claim)

(b) d.f. $= (r-1)(c-1) = 2$

$\chi_0^2 = 5.991$; Rejection region: $\chi^2 > 5.991$

(c)

| $O$ | $E$ | $O - E$ | $(O - E)^2$ | $\dfrac{(O - E)^2}{E}$ |
|---|---|---|---|---|
| 271 | 270.930 | 0.070 | 0.004900 | 0 |
| 257 | 257.286 | −0.286 | 0.081796 | 0.0003 |
| 149 | 148.784 | 0.216 | 0.046656 | 0.0003 |
| 146 | 146.070 | −0.070 | 0.004900 | 0 |
| 139 | 138.714 | 0.286 | 0.081796 | 0.0006 |
| 80 | 80.216 | −0.216 | 0.046656 | 0.0006 |
| | | | | 0.0018 |

$\chi^2 \approx 0.002$

(d) Because $\chi^2 < 5.991$, fail to reject $H_0$.

(e) There is not enough evidence at the 5% level of significance to conclude that the number of times former smokers tried to quit is dependent on gender.

**17.** (a) The claim is "the reason and the type of worker are dependent."

$H_0$ : Reasons are independent of the type of worker.

$H_a$ : Reasons are dependent on the type of worker. (claim)

(b) d.f. $= (r-1)(c-1) = 2$

$\chi_0^2 = 9.210$; Rejection region: $\chi^2 > 9.210$

(c)

| $O$ | $E$ | $O - E$ | $(O - E)^2$ | $\dfrac{(O - E)^2}{E}$ |
|---|---|---|---|---|
| 30 | 39.421 | −9.421 | 88.7552 | 2.2515 |
| 36 | 31.230 | 4.770 | 22.7529 | 0.7286 |
| 41 | 36.349 | 4.651 | 21.6318 | 0.5951 |
| 47 | 37.579 | 9.421 | 88.7552 | 2.3618 |
| 25 | 29.770 | −4.770 | 22.7529 | 0.7643 |
| 30 | 34.651 | −4.651 | 21.6318 | 0.6243 |
| | | | | 7.3256 |

$\chi^2 \approx 7.326$

(d) Because $\chi^2 < 9.210$, fail to reject $H_0$.

(e) There is not enough evidence at the 1% level of significance to conclude that reasons for continuing education are dependent on the type or worker.

**19.** (a) The claim is "the type of crash depends on the type of vehicle."

$H_0$ : The type of crash is independent of the type of vehicle.

$H_a$ : The type of crash is dependent on the type of vehicle. (claim)

(b) d.f. $= (r-1)(c-1) = 2$

$\chi_0^2 = 5.991$; Rejection region: $\chi^2 > 5.991$

(c)

| $O$ | $E$ | $O - E$ | $(O - E)^2$ | $\dfrac{(O - E)^2}{E}$ |
|------|-----------|-----------|-------------|-------------------------|
| 1163 | 1351.0258 | -188.0258 | 35,353.6868 | 26.1680 |
| 551 | 450.3419 | 100.6581 | 10,132.0490 | 22.4986 |
| 522 | 434.6323 | 87.3677 | 7633.1118 | 17.5622 |
| 1417 | 1228.9742 | 188.0258 | 35,353.6868 | 28.7668 |
| 309 | 409.6581 | -100.6581 | 10,132.0490 | 24.7329 |
| 308 | 395.3677 | -87.3677 | 7633.1118 | 19.3064 |
| | | | | 139.0349 |

$\chi^2 \approx 139.035$

(d) Because $\chi^2 > 5.991$, reject $H_0$.

(e) There is enough evidence at the 5% level of significance to conclude that the type of crash is dependent on the type of vehicle.

**21.** (a) The claim is "who borrows money for college in a family is related to the income of the family."

$H_0$ : Who borrows money for college in a family is independent of the family's income.

$H_a$ : Who borrows money for college in a family is dependent on the family's income. (claim)

(b) d.f. $= (r-1)(c-1) = 6$

$\chi_0^2 = 16.812$; Rejection region: $\chi^2 > 16.812$

(c)

| $O$ | $E$ | $O - E$ | $(O - E)^2$ | $\dfrac{(O - E)^2}{E}$ |
|------|----------|----------|-------------|-------------------------|
| 168 | 148.7213 | 19.2787 | 371.6686 | 2.4991 |
| 40 | 49.9826 | -9.9826 | 99.6527 | 1.9937 |
| 20 | 25.4513 | -5.4513 | 29.7164 | 1.1676 |
| 266 | 269.8448 | -3.8448 | 14.7826 | 0.0548 |
| 255 | 232.4146 | 22.5854 | 510.0981 | 2.1948 |
| 85 | 78.1105 | 6.8895 | 47.4653 | 0.6077 |
| 46 | 39.7741 | 6.2259 | 38.7624 | 0.9746 |
| 386 | 421.7008 | -35.7009 | 1274.5476 | 3.0224 |
| 62 | 103.8641 | -41.8641 | 1752.5995 | 16.8740 |
| 38 | 34.9069 | 3.0931 | 9.5673 | 0.2741 |
| 17 | 17.7747 | -0.7747 | 0.6001 | 0.0338 |
| 228 | 188.4544 | 39.5456 | 1563.8564 | 8.2983 |
| | | | | 37.9948 |

$\chi^2 \approx 37.99$

(d) Because $\chi^2 > 16.812$, reject $H_0$.

(e) There is enough evidence at the 1% level of significance to conclude that who borrows money for college in a family is dependent on the family's income.

23. The claim is "the proportions of motor vehicle crash deaths involving males and females are the same for each group."

$H_0$ : The proportions are equal. (claim)

$H_a$ : At least one of proportion is different from the others.

d.f. $= (r-1)(c-1) = 7$

$\chi_0^2 = 14.067$ ; Rejection region: $\chi^2 > 14.067$

| $O$ | $E$ | $O - E$ | $(O - E)^2$ | $\dfrac{(O - E)^2}{E}$ |
|---|---|---|---|---|
| 110 | 109.0394 | 0.9606 | 0.9227 | 0.0085 |
| 94 | 89.7972 | 4.2028 | 17.6637 | 0.1967 |
| 73 | 71.2676 | 1.7324 | 3.0012 | 0.0421 |
| 87 | 83.3831 | 3.6169 | 13.0820 | 0.1569 |
| 68 | 64.8535 | 3.1465 | 9.9003 | 0.1527 |
| 37 | 41.3352 | -4.3352 | 18.7941 | 0.4547 |
| 26 | 31.3577 | -5.3577 | 28.7054 | 0.1527 |
| 11 | 14.9662 | -3.9662 | 15.7307 | 1.0511 |
| 43 | 43.9606 | -0.9606 | 0.9227 | 0.0210 |
| 32 | 36.2028 | -4.2028 | 17.6637 | 0.4879 |
| 27 | 28.7324 | -1.7324 | 3.0012 | 0.1045 |
| 30 | 33.6169 | -3.6169 | 13.0820 | 0.3891 |
| 23 | 26.1465 | -3.1465 | 9.9003 | 0.3786 |
| 21 | 16.6648 | 4.3352 | 18.7941 | 1.1278 |
| 18 | 12.6423 | 5.3577 | 28.7054 | 0.3786 |
| 10 | 6.0338 | 3.9662 | 15.7307 | 2.6071 |
| | | | | 7.7099 |

$\chi^2 \approx 7.710$

Because $\chi^2 < 14.067$ , fail to reject $H_0$ . There is not enough evidence at the 5% level of significance to reject the claim that the proportions of motor vehicle crash deaths involving males and females are the same for each group.

25. Right-tailed

27.

| Status | Educational attainment | | | |
|---|---|---|---|---|
| | Not a high school graduate | High school graduate | Some college, no degree | Associate's, bachelor's, or advanced degree |
| Employed | 0.049 | 0.171 | 0.103 | 0.283 |
| Unemployed | 0.009 | 0.023 | 0.011 | 0.016 |
| Not in the labor force | 0.070 | 0.119 | 0.054 | 0.091 |

29. Several of the expected frequencies are less than 5.

**31.** 46.7%

**33.**

| Status | Educational attainment | | | |
|---|---|---|---|---|
| | Not a high school graduate | High school graduate | Some college, no degree | Associate's, bachelor's, or advanced degree |
| Employed | 0.384 | 0.546 | 0.613 | 0.725 |
| Unemployed | 0.074 | 0.073 | 0.065 | 0.041 |
| Not in the labor force | 0.543 | 0.380 | 0.321 | 0.234 |

**35.** 7.4%

## 10.3 COMPARING TWO VARIANCES

## 10.3 Try It Yourself Solutions

**1a.** $\alpha = 0.05$     **b.** $F_0 = 2.45$

**2a.** $\alpha = 0.01$     **b.** $F_0 = 18.31$

**3a.** The claim is "the variance of the time required for nutrients to enter the bloodstream is less with the specially treated intravenous solution than the variance of the time without the solution."

$H_0 : \sigma_1^2 \leq \sigma_2^2$; $H_a : \sigma_1^2 > \sigma_2^2$ (claim)

**b.** $\alpha = 0.01$

**c.** $\text{d.f.}_N = n_1 - 1 = 24$

$\text{d.f.}_D = n_2 - 1 = 19$

**d.** $F_0 = 2.92$; Rejection region: $F > 2.92$

**e.** $F = \dfrac{s_1^2}{s_2^2} = \dfrac{180}{56} \approx 3.21$

**f.** Because $F > 2.92$, reject $H_0$.

**g.** There is enough evidence at the 1% level of significance to support the researcher's claim that a specially treated intravenous solution decreases the variance of the time required for nutrients to enter the bloodstream.

**4a.** The claim is "the pH levels of the soil in two geographic regions have equal standard deviations."

$H_0 : \sigma_1 = \sigma_2$ (claim); $H_a : \sigma_1 \neq \sigma_2$

**b.** $\alpha = .01$

**c.** $\text{d.f.}_N = n_1 - 1 = 15$

$\text{d.f.}_D = n_2 - 1 = 21$

**d.** $F_0 = 3.43$; Rejection region: $F > 3.43$

**e.** $F = \dfrac{s_1^2}{s_2^2} = \dfrac{(0.95)^2}{(0.78)^2} \approx 1.48$

**f.** Because $F < 3.43$. fail to reject $H_0$.

**g.** There is not enough evidence at the 1% level of significance to reject the biologist's claim that the pH levels of the soil in the two geographic regions have equal standard deviations.

## 10.3 EXERCISE SOLUTIONS

1.  Specify the level of significance $\alpha$. Determine the degrees of freedom for the numerator and denominator. Use Table 7 in Appendix B to find the critical value $F$.

3.  (1) The samples must be randomly selected, (2) the samples must be independent, and (3) each population must have a normal distribution.

5.  $F_0 = 2.54$          7.  $F_0 = 2.06$          9.  $F_0 = 9.16$          11.  $F_0 = 1.80$

13.  $H_0 : \sigma_1^2 \le \sigma_2^2$; $H_a : \sigma_1^2 > \sigma_2^2$ (claim)

  $\text{d.f.}_N = 4$,  $\text{d.f.}_D = 5$

  $F_0 = 3.52$;  Rejection region: $F > 3.52$

  $F = \dfrac{s_1^2}{s_2^2} = \dfrac{773}{765} \approx 1.010$

  Because $F < 3.52$, fail to reject $H_0$. There is not enough evidence at the 10% level of significance to support the claim.

15.  $H_0 : \sigma_1^2 \le \sigma_2^2$ (claim); $H_a : \sigma_1^2 > \sigma_2^2$

  $\text{d.f.}_N = 10$,  $\text{d.f.}_D = 9$

  $F_0 = 5.26$;  Rejection region: $F > 5.26$

  $F = \dfrac{s_1^2}{s_2^2} = \dfrac{842}{836} \approx 1.007$

  Because $F < 5.26$, fail to reject $H_0$. There is not enough evidence at the 1% level of significance to reject the claim.

17.  $H_0 : \sigma_1^2 = \sigma_2^2$ (claim); $H_a : \sigma_1^2 \ne \sigma_2^2$

  $\text{d.f.}_N = 12$,  $\text{d.f.}_D = 19$

  $F_0 = 3.76$;  Rejection region: $F > 3.76$

  $F = \dfrac{s_1^2}{s_2^2} = \dfrac{9.8}{2.5} \approx 3.920$

  Because $F > 3.76$, reject $H_0$. There is enough evidence at the 1% level of significance to reject the claim.

**19.** Population 1: Company B; Population 2: Company A

    (a) The claim is "the variance of the life of Company A's appliances is less than the variance of the life of Company B's appliances."

$$H_0: \sigma_1^2 \leq \sigma_2^2; \ H_a: \sigma_1^2 > \sigma_2^2 \text{ (claim)}$$

    (b) d.f.$_N = 24$, d.f.$_D = 19$

$$F_0 = 2.11; \text{ Rejection region: } F > 2.11$$

    (c) $F = \dfrac{s_1^2}{s_2^2} = \dfrac{3.9}{1.8} \approx 2.167$

    (d) Because $F > 2.11$, reject $H_0$.

    (e) There is enough evidence at the 5% level of significance to support Company A's claim that the variance of life of its appliances is less than the variance of life of Company B appliances.

**21.** Population 1: Age group 35-49; Population 2: Age group 18-34

    (a) The claim is "the variances of the waiting times differ between the two age groups."

$$H_0: \sigma_1^2 = \sigma_2^2; \ H_a: \sigma_1^2 \neq \sigma_2^2 \text{ (claim)}$$

    (b) d.f.$_N = 6$, d.f.$_D = 4$

$$F_0 = 9.20; \text{ Rejection region: } F > 9.20$$

    (c) $s_1^2 = 63.48$, $s_2^2 = 38.3$

$$F = \dfrac{s_1^2}{s_2^2} = \dfrac{63.48}{38.3} \approx 1.66$$

    (d) Because $F < 9.20$, fail to reject $H_0$.

    (e.) There is not enough evidence at the 5% level of significance to conclude that the variances of the waiting times differ between the two age groups.

**23.** Population 1: District 1; Population 2: District 2

    (a) The claim is "the standard deviations of science achievement test scores for eighth grade students are the same in Districts 1 and 2."

$$H_0: \sigma_1^2 = \sigma_2^2 \text{(claim)}; \ H_a: \sigma_1^2 \neq \sigma_2^2$$

    (b) d.f.$_N = 11$, d.f.$_D = 13$

$$F_0 = 2.635; \text{ Rejection region: } F > 2.635$$

    (c) $F = \dfrac{s_1^2}{s_2^2} = \dfrac{(36.8)^2}{(32.5)^2} \approx 1.282$

    (d) Because $F < 2.635$, fail to reject $H_0$.

    (e) There is not enough evidence at the 10% level of significance to reject the administrator's claim that the standard deviations of science assessment test scores for eighth grade students are the same in Districts 1 and 2.

**25.** Population 1: New York; Population 2: California

    (a) The claim is "the standard deviation of the annual salaries for actuaries is greater in New York than in California."

$$H_0: \sigma_1^2 \leq \sigma_2^2; \ H_a: \sigma_1^2 > \sigma_2^2 \text{ (claim)}$$

    (b) d.f.$_N = 40$, d.f.$_D = 60$

$$F_0 = 1.59; \text{ Rejection region: } F > 1.59$$

(c) $F = \dfrac{s_1^2}{s_2^2} = \dfrac{(39,700)^2}{(29,000)^2} \approx 1.87$

(d) Because $F > 1.59$, reject $H_0$.

(e) There is enough evidence at the 5% level of significance to conclude the standard deviation of the annual salaries for actuaries is greater in New York than in California.

27. Right-tailed: $F_R = 14.73$

Left-tailed:

(1) d.f.$_N = 3$ and d.f.$_D = 6$

(2) $F = 6.60$

(3) Critical value is $\dfrac{1}{F} = \dfrac{1}{6.60} \approx 0.15$.

29. $\dfrac{s_1^2}{s_2^2}\dfrac{1}{F_R} < \dfrac{\sigma_1^2}{\sigma_2^2} < \dfrac{s_1^2}{s_2^2}\dfrac{1}{F_L} \rightarrow \left(\dfrac{10.89}{9.61}\right)\left(\dfrac{1}{3.33}\right) < \dfrac{\sigma_1^2}{\sigma_2^2} < \left(\dfrac{10.89}{9.61}\right)(3.02) \rightarrow 0.340 < \dfrac{\sigma_1^2}{\sigma_2^2} < 3.422$

## 10.4 ANALYSIS OF VARIANCE

## 10.4 Try It Yourself Solutions

**1a.** The claim is "there is a difference in the mean a monthly sales among the sales regions."

$H_0 : \mu_1 = \mu_2 = \mu_3 = \mu_4$

$H_a$ : At least one mean is different from the others. (claim)

**b.** $\alpha = 0.05$

**c.** d.f.$_N = 3$;  d.f.$_D = 14$

**d.** $F_0 = 3.34$; Rejection region: $F > 3.34$

**e.**

| Variation | Sum of Squares | Degrees of Freedom | Mean Squares | $F$ |
|---|---|---|---|---|
| Between | 549.8 | 3 | 183.3 | 4.22 |
| Within | 608.0 | 14 | 43.4 | |

$F \approx 4.22$

**f.** Because $F > 3.34$, reject $H_0$.

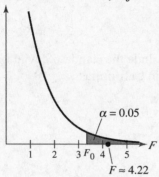

$\alpha = 0.05$

$F \approx 4.22$

**g.** There is enough evidence at the 5% level of significance to conclude that there is a difference in the mean monthly sales among the sales regions.

**2a.** The claim is "there is a difference in the means of the GPAs."

$H_0 : \mu_1 = \mu_2 = \mu_3 = \mu_4$

$H_a :$ At least one mean is different from the others. (claim)

**b.** Enter the data.

**c.**

| Variation | Sum of Squares | Degrees of Freedom | Mean Squares | $F$ |
|---|---|---|---|---|
| Between | 0.584 | 3 | 0.195 | 1.34 |
| Within | 4.360 | 30 | 0.145 | |

$F = 1.34 \rightarrow P\text{-value} = 0.280$

**c.** $0.280 > 0.05$

**d.** Because $P\text{-value} > 0.05$, fail to reject $H_0$. There is not enough evidence to conclude that there is a difference in the means of the GPAs.

## 10.4 EXERCISE SOLUTIONS

**1.** $H_0 : \mu_1 = \mu_2 = \ldots = \mu_k$

$H_a :$ At least one mean is different from the others.

**3.** The $MS_B$ measures the differences related to the treatment given to each sample.

The $MS_W$ measures the differences related to entries within the same sample.

**5.** (a) The claim is that "the mean costs per ounce are different."

$H_0 : \mu_1 = \mu_2 = \mu_3$

$H_a :$ At least one mean is different from the others. (claim)

(b) $\text{d.f.}_N = k - 1 = 2$; $\text{d.f.}_D = N - k = 12$

$F_0 = 3.89$; Rejection region: $F > 3.89$

(c)

| Variation | Sum of Squares | Degrees of Freedom | Mean Squares | $F$ |
|---|---|---|---|---|
| Between | 0.07406 | 2 | 0.037032 | 4.80 |
| Within | 0.09267 | 12 | 0.007723 | |

$F \approx 4.80$

(d) Because $F > 3.89$, reject $H_0$.

(e) There is enough evidence at the 5% level of significance to conclude that at least one mean cost per ounce is different from the others.

7. (a) The claim is "at least one mean vacuum cleaner weight is different from the others."

$H_0 : \mu_1 = \mu_2 = \mu_3$

$H_a$ : At least one mean is different from the others. (claim)

(b) d.f.$_N = k - 1 = 2$; d.f.$_D = N - k = 15$

$F_0 = 6.36$; Rejection region: $F > 6.36$

(c)

| Variation | Sum of Squares | Degrees of Freedom | Mean Squares | F |
|---|---|---|---|---|
| Between | 100.3333 | 2 | 50.16667 | 12.10 |
| Within | 62.16667 | 15 | 4.14444 | |

$F \approx 12.10$

(d) Because $F > 6.36$, reject $H_0$.

(e) There is enough evidence at the 1% level of significance to conclude that at least one mean vacuum cleaner weight is different from the others.

9. (a) The claim is "at least one mean age is different from the others."

$H_0 : \mu_1 = \mu_2 = \mu_3 = \mu_4$

$H_a$ : At least one mean is different from the others. (claim)

(b) d.f.$_N = k - 1 = 3$; d.f.$_D = N - k = 48$

$F_0 = 2.84$; Rejection region: $F > 2.84$

(c)

| Variation | Sum of Squares | Degrees of Freedom | Mean Squares | F |
|---|---|---|---|---|
| Between | 10.07692 | 3 | 3.358974 | 0.62 |
| Within | 262.1538 | 48 | 5.461538 | |

$F \approx 0.62$

(d) Because $F < 2.84$, fail to reject $H_0$.

(e) There is not enough evidence at the 5% level of significance to conclude that at least one mean age is different from the others.

11. (a) The claim is "the mean scores are the same for all regions."

$H_0 : \mu_1 = \mu_2 = \mu_3 = \mu_4$ (claim)

$H_a$ : At least one mean is different from the others.

(b) d.f.$_N = k - 1 = 3$; d.f.$_D = N - k = 30$

$F_0 = 2.28$; Rejection region: $F > 2.28$

(c)

| Variation | Sum of Squares | Degrees of Freedom | Mean Squares | F |
|---|---|---|---|---|
| Between | 40.99123 | 3 | 13.66374 | 7.49 |
| Within | 54.71936 | 30 | 1.823979 | |

$F \approx 7.49$

(d) Because $F > 2.28$, reject $H_0$.

(e) There is enough evidence at the 10% level of significance to reject the claim that the mean scores are the same for all regions.

13. (a) The claim is "the mean salary is different in at least one of the areas."

$H_0 : \mu_1 = \mu_2 = \mu_3 = \mu_4 = \mu_5 = \mu_6$

$H_a$ : At least one mean is different from the others. (claim)

(b) $\text{d.f.}_N = k - 1 = 5$;   $\text{d.f.}_D = N - k = 30$

$F_0 = 2.53$; Rejection region: $F > 2.53$

(c)

| Variation | Sum of Squares | Degrees of Freedom | Mean Squares | F |
|---|---|---|---|---|
| Between | 370,963,606 | 5 | 74,192,721 | 2.28 |
| Within | 975,795,651 | 30 | 32,526,522 | |

$F \approx 2.28$

(d) Because $F < 2.53$, fail to reject $H_0$.

(e) There is not enough evidence at the 5% level of significance to conclude that the mean salary is different in at least one of the areas.

15. $H_0$ : Advertising medium has to effect on mean ratings.

$H_a$ : Advertising medium has an effect on mean ratings.

$H_0$ : Length of ad has no effect on mean ratings.

$H_a$ : Length of ad has an effect on mean ratings.

$H_0$ : There is no interaction effect between advertising medium and length of ad on mean ratings.

$H_a$ : There is an interaction effect between advertising medium and length of ad on mean ratings.

| Source | d.f. | SS | MS | F | P |
|---|---|---|---|---|---|
| Ad medium | 1 | 1.25 | 1.25 | 0.57 | 0.459 |
| Length of ad | 1 | 0.45 | 0.45 | 0.21 | 0.655 |
| Interaction | 1 | 0.45 | 0.45 | 0.21 | 0.655 |
| Error | 16 | 34.80 | 2.175 | | |
| Total | 19 | 36.95 | | | |

Fail to reject all null hypotheses. The interaction between the advertising medium and length of the ad has no effect on the rating and therefore there is no significant difference in the means of the ratings.

17. $H_0$ :  Age has no effect on mean GPA.

$H_a$ : Age has an effect on mean GPA.

$H_0$ : Gender has no effect on mean GPA.

$H_a$ : Gender has an effect on mean GPA.

$H_0$ : There is no interaction effect between age and gender on mean GPA.

$H_a$ : There is no interaction effect between age and gender on mean GPA.

| Source | d.f. | SS | MS | F | P |
|---|---|---|---|---|---|
| Age | 3 | 0.4146 | 0.138 | 0.12 | 0.948 |
| Gender | 1 | 0.1838 | 0.184 | 0.16 | 0.697 |
| Interaction | 3 | 0.2912 | 0.097 | 0.08 | 0.968 |
| Error | 16 | 18.6600 | 1.166 | | |
| Total | 23 | 19.5496 | | | |

Fail to reject all null hypotheses. The interaction between age and gender has no effect GPA and therefore there is no significant difference in the means of the GPAs.

**19.**

| | Mean | Size |
|---|---|---|
| Pop 1 | 0.455 | 6 |
| Pop2 | 0.606 | 5 |
| Pop3 | 0.460 | 4 |

$SS_W = 0.09267$

$\sum (n_i - 1) = N - k = 12$

$F_0 = 3.89 \rightarrow CV_{\text{Scheffe}'} = 3.89(3-1) = 7.78$

$$\frac{(\overline{x}_1 - \overline{x}_2)^2}{\dfrac{SS_W}{\sum (n_i - 1)} \left[ \dfrac{1}{n_1} + \dfrac{1}{n_2} \right]} \approx 8.05 \rightarrow \text{Significant difference}$$

$$\frac{(\overline{x}_1 - \overline{x}_3)^2}{\dfrac{SS_W}{\sum (n_i - 1)} \left[ \dfrac{1}{n_1} + \dfrac{1}{n_3} \right]} \approx 0.01 \rightarrow \text{No difference}$$

$$\frac{(\overline{x}_2 - \overline{x}_3)^2}{\dfrac{SS_W}{\sum (n_i - 1)} \left[ \dfrac{1}{n_2} + \dfrac{1}{n_3} \right]} \approx 6.13 \rightarrow \text{No difference}$$

**21.**

| | Mean | Size |
|---|---|---|
| Pop 1 | 73.06 | 10 |
| Pop2 | 50.82 | 10 |
| Pop3 | 43.97 | 10 |

$SS_W = 1918.441$

$\sum (n_i - 1) = N - k = 27$

$F_0 = 5.49 \rightarrow CV_{\text{Scheffe}'} = 5.49(3-1) = 10.98$

$$\frac{(\overline{x}_1 - \overline{x}_2)^2}{\dfrac{SS_W}{\sum (n_i - 1)} \left[ \dfrac{1}{n_1} + \dfrac{1}{n_2} \right]} \approx 34.81 \rightarrow \text{Significant difference}$$

$$\frac{(\overline{x}_1 - \overline{x}_3)^2}{\dfrac{SS_W}{\sum(n_i - 1)}\left[\dfrac{1}{n_1} + \dfrac{1}{n_3}\right]} \approx 59.55 \to \text{Significant difference}$$

$$\frac{(\overline{x}_1 - \overline{x}_4)^2}{\dfrac{SS_W}{\sum(n_i - 1)}\left[\dfrac{1}{n_1} + \dfrac{1}{n_4}\right]} \approx 3.30 \to \text{No difference}$$

# CHAPTER 10 REVIEW EXERCISE SOLUTIONS

1.  (a)

| Response | Distribution |
|----------|--------------|
| Less than $10 | 29% |
| $10 to $20 | 16% |
| More than $21 | 9% |
| Don't give one/other | 46% |

$H_0$ : The distribution of the allowance amounts is 29% less than $10, 16% $10 to $20, 9% more than $21, and 46% don't give one/other.

$H_a$ : The distribution of amounts differ from the claimed or expected distribution. (claim)

(b) $\chi_0^2 = 6.251$ ; Rejection region: $\chi^2 > 6.251$

(c)

| Response | Distribution | Observed | Expected | $\dfrac{(O - E)^2}{E}$ |
|----------|--------------|----------|----------|------------------------|
| Less than $10 | 29% | 353 | 319.87 | 3.431 |
| $10 to $20 | 16% | 167 | 176.48 | 0.509 |
| More than $21 | 9% | 94 | 99.27 | 0.280 |
| Don't give one/other | 46% | 489 | 507.38 | 0.666 |
|  |  |  |  | 4.886 |

$\chi^2 \approx 4.886$

(d) Because $\chi^2 < 6.251$, fail to reject $H_0$.

(e) There is not enough evidence at the 10% level of significance to conclude that there has been a change in the claimed or expected distribution.

3.  (a)

| Response | Distribution |
|----------|--------------|
| Short-game shots | 65% |
| Approach and swing | 22% |
| Driver shots | 9% |
| Putting | 4% |

$H_0$ : The distribution of responses from golf students about what they need the most help with is 22% approach and swing, 9% driver shots, 4% putting and 65% short-game shots. (claim)

$H_a$ : The distribution of responses differs from the claimed or expected distribution.

(b) $\chi_0^2 = 7.815$; Rejection region: $\chi^2 > 7.815$

(c)

| Response | Distribution | Observed | Expected | $\dfrac{(O-E)^2}{E}$ |
|---|---|---|---|---|
| Short-game shots | 65% | 276 | 282.75 | 0.161 |
| Approach and swing | 22% | 99 | 95.70 | 0.114 |
| Driver shots | 9% | 42 | 39.15 | 0.207 |
| Putting | 4% | 18 | 17.40 | 0.021 |
| | | | | 0.503 |

$\chi^2 \approx 0.503$

(d) Because $\chi^2 < 7.815$, fail to reject $H_0$.

(e) There is not enough evidence at the 5% level of significance to conclude that the distribution of golf students' responses is the same as the claimed or expected distribution.

5. (a) Expected frequencies:

| Gender | Years of full-time teaching experience | | | | |
|---|---|---|---|---|---|
| | Less than 3 years | 3–9 years | 10–20 years | 20 years or more | Total |
| Male | 141.3 | 352.2 | 285.6 | 270.9 | 1050 |
| Female | 329.7 | 821.8 | 666.4 | 632.1 | 2450 |
| Total | 471 | 1174 | 952 | 903 | 3500 |

(b) The claim is "gender is related to the years of full-time teaching experience."

$H_0$ : Years of full-time teaching experience is independent of gender.

$H_a$ : Years of full-time teaching experience is dependent on gender. (claim)

(c) d.f. $= (r-1)(c-1) = 3$; $\chi_0^2 = 11.345$; Rejection region: $\chi^2 > 11.345$

(d)

| $O$ | $E$ | $O - E$ | $(O - E)^2$ | $\dfrac{(O-E)^2}{E}$ |
|---|---|---|---|---|
| 143 | 141.3 | -1.7 | 2.89 | 0.0205 |
| 349 | 352.2 | 3.2 | 10.24 | 0.0291 |
| 279 | 285.6 | 6.6 | 43.56 | 0.1525 |
| 279 | 270.9 | -8.1 | 65.61 | 0.2422 |
| 328 | 329.7 | 1.7 | 2.89 | 0.0088 |
| 825 | 821.8 | -3.2 | 10.24 | 0.0125 |
| 673 | 666.4 | -6.6 | 43.56 | 0.0654 |
| 624 | 632.1 | 8.1 | 65.61 | 0.1038 |
| | | | | 0.635 |

$\chi^2 \approx 0.635$

(e) Because $\chi^2 < 11.345$, fail to reject $H_0$.

(f) There is not enough evidence at the 1% level of significance to conclude that the years of full-time teaching experience is dependent on gender.

**7.** (a) Expected frequencies:

| Status | Mammals | Birds | Reptiles | Amphibians | Fish | Total |
|---|---|---|---|---|---|---|
| | | | **Vertebrae Group** | | | |
| Endangered | 145.8 | 128.79 | 50.22 | 13.77 | 66.42 | 405 |
| Threatened | 34.2 | 30.21 | 11.78 | 3.23 | 15.58 | 95 |
| Total | 180 | 159 | 62 | 17 | 82 | 500 |

(b) The claim is "the species' status (endangered or threatened) is independent of vertebrate group."

$H_0$ : Species' status (endangered or threatened) is independent of vertebrate group. (claim)

$H_a$ : Species' status (endangered or threatened) is dependent on vertebrate group.

(c) d.f. $= (r-1)(c-1) = 4$; $\chi_0^2 = 13.277$; Rejection region: $\chi^2 > 13.277$

(d)

| $O$ | $E$ | $O - E$ | $(O - E)^2$ | $\dfrac{(O - E)^2}{E}$ |
|---|---|---|---|---|
| 162 | 145.8 | -16.2 | 262.44 | 1.800 |
| 143 | 128.79 | -14.21 | 201.9241 | 1.568 |
| 41 | 50.22 | 9.22 | 85.0084 | 1.693 |
| 12 | 13.77 | 1.77 | 3.1329 | 0.228 |
| 47 | 66.42 | 19.42 | 377.1364 | 5.678 |
| 18 | 34.2 | 16.2 | 262.44 | 7.674 |
| 16 | 30.21 | 14.21 | 201.9241 | 6.684 |
| 21 | 11.78 | -9.22 | 85.0084 | 7.216 |
| 5 | 3.23 | -1.77 | 3.1329 | 0.970 |
| 35 | 15.58 | -19.42 | 377.1364 | 24.206 |
| | | | | 57.717 |

$\chi^2 \approx 57.717$

(e) Because $\chi^2 > 13.277$, reject $H_0$.

(c) There is enough evidence at the 1% level of significance to reject the claim that species' status (endangered or threatened) is independent on vertebrate group.

**9.** $F_0 \approx 2.295$     **11.** $F_0 = 2.39$     **13.** $F_0 = 2.06$     **15.** $F_0 = 2.08$

**17.** (a)     The claim is "the variation in wheat production is greater in Garfield County than in Kay County."

$H_0 : \sigma_1^2 \le \sigma_2^2$; $H_a : \sigma_1^2 > \sigma_2^2$ (claim)

(b) d.f.$_N = 20$;  d.f.$_D = 15$;

$F_0 = 1.92$; Rejection region: $F > 1.92$

(c) $F = \dfrac{s_1^2}{s_2^2} = \dfrac{(0.76)^2}{(0.58)^2} \approx 1.72$

(d) Because $F < 1.92$, fail to reject $H_0$.

(e) There is not enough evidence at the 10% level of significance to support the claim that the variation in wheat production is greater in Garfield County than in Kay County.

**19.** (a) The claim is "the variance of SAT critical reading scores for females is different than the variance of SAT critical reading scores for males."

$H_0 : \sigma_1^2 = \sigma_2^2$; $H_a : \sigma_1^2 \ne \sigma_2^2$ (claim)

(b) $\text{d.f.}_N = 12$; $\text{d.f.}_D = 8$

$F_0 = 7.01$; Rejection region: $F > 7.01$

(c) $F = \dfrac{s_1^2}{s_2^2} = \dfrac{20,002.56}{17,136.11} \approx 1.17$

(d) Because $F < 7.01$, fail to reject $H_0$.

(e) There is not enough evidence at the 1% level of significance to support the claim that the test score variance for females is different than the test score variance for males.

21. (a) The claim is "the mean amount spent on energy in one year is different in at least one of the regions."

$H_0 : \mu_1 = \mu_2 = \mu_3 = \mu_4$

$H_a$ : At least one mean is different from the others. (claim)

(b) $\text{d.f.}_N = k - 1 = 3$; $\text{d.f.}_D = N - k = 28$

$F_0 = 2.29$; Rejection region: $F > 2.29$

(c)

| Variation | Sum of Squares | Degrees of Freedom | Mean Squares | F |
|---|---|---|---|---|
| Between | 2,207,334.3 | 3 | 735,778.08 | 6.19 |
| Within | 3,329,531.8 | 28 | 118,911.85 | |

$F \approx 6.19$

(d) Because $F > 2.29$, reject $H_0$.

(e) There is enough evidence at the 10% level of significance to conclude that at least one mean amount spent on energy is different from the others.

## CHAPTER 10 QUIZ SOLUTIONS

1. (a) $H_0$ : The distribution of educational achievement for people in the United States ages 30–34 is 12.4% not a high school graduate, 30.4% high school graduate, 16.7% some college, no degree; 9.6% associate's degree, 19.8% bachelor's degree, and 11.1% advanced degree.

$H_a$ : The distribution of educational achievement for people in the United States ages 30–34 differs from the claimed distribution. (claim)

(b) $\chi_0^2 = 11.071$; Rejection region: $\chi^2 > 11.071$

(c)

| Distribution | | Observed | Expected | $\dfrac{(O-E)^2}{E}$ |
|---|---|---|---|---|
| Not a H.S. graduate | 12.4% | 36 | 39.804 | 0.3635 |
| H.S. graduate | 30.4% | 84 | 97.584 | 1.8909 |
| Some college no degree | 16.7% | 56 | 53.607 | 0.1068 |
| Associate's degree | 9.6% | 34 | 30.816 | 0.3290 |
| Bachelor's degree | 19.8% | 73 | 63.558 | 1.4027 |
| Advanced degree | 11.1% | 38 | 35.631 | 0.1575 |
| | | 321 | | 4.2505 |

$\chi^2 \approx 4.25$

(d) Because $\chi^2 < 11.071$, fail to reject $H_0$.

(e) There is not enough evidence at the 5% level of significance to conclude that the distribution for people in the United States ages 30–34 differs from the distribution for people ages 25 and older.

2. (a) $H_0$ : Age and educational attainment are independent.

$H_a$ : Age and educational attainment are dependent. (claim)

(b) d.f. $= (r-1)(c-1) = 5$; $\chi_0^2 = 15.086$; Rejection region: $\chi^2 > 15.086$

(c)

| $O$ | $E$ | $O-E$ | $(O-E)^2$ | $\dfrac{(O-E)^2}{E}$ |
|------|---------|----------|----------|--------|
| 36 | 40.7487 | 4.7487 | 22.5505 | 0.5534 |
| 84 | 96.4663 | 12.4663 | 155.4092 | 1.6110 |
| 56 | 53.6386 | -2.3614 | 5.5762 | 0.1040 |
| 34 | 29.1062 | -4.8938 | 23.9491 | 0.8228 |
| 73 | 60.7073 | -12.2927 | 151.1116 | 2.4892 |
| 38 | 40.3329 | 2.3329 | 5.4424 | 0.1349 |
| 62 | 57.2513 | -4.7487 | 22.5505 | 0.3939 |
| 148 | 135.5337 | -12.4663 | 155.4092 | 1.1466 |
| 73 | 75.3614 | 2.3614 | 5.5762 | 0.0740 |
| 36 | 40.8938 | 4.8938 | 23.9491 | 0.5856 |
| 73 | 85.2928 | 12.2927 | 151.1116 | 1.7717 |
| 59 | 56.6671 | -2.3329 | 5.4424 | 0.0960 |
| | | | | 9.7832 |

$\chi^2 \approx 9.783$

(d) Because $\chi^2 < 15.086$, fail to reject $H_0$.

(e) There is not enough evidence at the 1% level of significance to conclude that educational attainment is dependent on age.

3. Population 1: Ithaca $\rightarrow s_1^2 = 99.466$

Population 2: Little Rock $\rightarrow s_2^2 = 73.576$

(a) $H_0 : \sigma_1^2 = \sigma_2^2$; $H_a : \sigma_1^2 \neq \sigma_2^2$ (claim)

(b) d.f.$_N = 12$, d.f.$_D = 14$;

$F_0 = 4.43$; Rejection region: $F > 4.43$

(c) $F = \dfrac{s_1^2}{s_2^2} = \dfrac{99.466}{73.876} \approx 1.35$

(d) Because $F < 4.43$, fail to reject $H_0$.

(e) There is not enough evidence at the 1% level of significance to conclude the variances in annual wages for Ithaca, NY and Little Rock, AR are different.

4. (a) $H_0 : \mu_1 = \mu_2 = \mu_3$ (claim)

$H_a$ : At least one mean is different from the others.

(b) d.f.$_N = 2$, d.f.$_D = 41$

$F_0 = 2.44$; Rejection region: $F > 2.44$

(c)

| Variation | Sum of Squares | Degrees of Freedom | Mean Squares | F |
|---|---|---|---|---|
| Between | 710.6286 | 2 | 355.3143 | 4.52 |
| Within | 3145.0774 | 41 | 78.6269 | |

$F \approx 4.52$

(d) Because $F > 2.44$, reject $H_o$.

(e) There is enough evidence at the 10% level of significance to reject the claim that the mean annual wages are equal for all three cities.

## CUMULATIVE REVIEW, CHAPTERS 9-10

1.  (a)

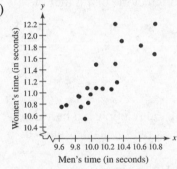

Men's time (in seconds)

$r \approx 0.823$; There is a strong, positive linear correlation.

(b) There is enough evidence at the 5% level of significance to conclude that there is a significant linear correlation between the men's and women's winning 100-meter times.

(c) $y = 1.225x - 1.181$

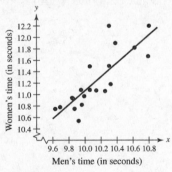

Men's time (in seconds)

(d) 10.95 seconds

3.  (a) $\hat{y} = 11,509 + 0.139(110,000) - 0.069(100,000) = 19,899$ pounds per acre

(b) $\hat{y} = 11,509 + 0.139(125,000) - 0.069(115,000) = 20,949$ pounds per acre

5.  $H_0$: The distribution of credit card debts for college students are distributed as 32.8% $0, 42.2% $1-$500, 11.4% $501-$1000, 5.1% $1001-$2000, 5.3% $2001-$4000, and 3.2% More than $4000. (claim)

$H_a$: The distribution of credit card debts for college students differs from the claimed distribution.

$\chi_0^2 = 11.071$; Rejection region: $\chi^2 > 11.071$

| | Distribution | Observed | Expected | $\dfrac{(O-E)^2}{E}$ |
|---|---|---|---|---|
| $0 | .328 | 290 | 295.2 | 0.0916 |
| $1-$500 | .422 | 397 | 379.8 | 0.7789 |
| $501-$1000 | .114 | 97 | 102.6 | 0.3057 |
| $1001-$2000 | .051 | 54 | 45.9 | 1.4294 |
| $2001-$4000 | .053 | 40 | 47.7 | 1.2430 |
| More than $4000 | .032 | 22 | 28.8 | 1.6056 |
| | | | | 5.4541 |

$\chi^2 \approx 5.45$

Because $\chi^2 < 11.071$, fail to reject $H_0$.

There is not enough evidence at the 5% level of significance to reject the claim that the distributions are the same.

7. (a) $r^2 = \dfrac{\sum\left(\hat{y}_i - \overline{y}\right)^2}{\sum\left(y_i - \overline{y}\right)^2} \approx 0.733$

About 73.3% of the variation in height can be explained by the relationship between metacarpal bone length and height, and about 26.7% of the variation is unexplained.

(b) $s_e = \sqrt{\dfrac{\sum\left(y_i - \hat{y}_i\right)^2}{n-2}} = \sqrt{\dfrac{126.7121}{7}} \approx 4.255$

The standard error of estimate of the height for a specific metacarpal bone length is about 4.255 centimeters.

(c) $n = 9$ d.f. $= 7$, $t_c = 2.365$, $s_e \approx 4.255$

$\hat{y} = 1.70x + 94.43 = 1.70(50) + 94.43 = 179.43$

$E = t_c s_e \sqrt{1 + \dfrac{1}{n} + \dfrac{n\left(x - \overline{x}\right)^2}{n\left(\sum x^2\right) - \left(\sum x\right)^2}}$

$\approx (2.365)(4.255)\sqrt{1 + \dfrac{1}{9} + \dfrac{9\left(50 - 409/9\right)^2}{9(18,707) - (409)^2}}$

$\approx 11.401$

$\hat{y} \pm E \rightarrow 179.43 \pm 11.401 \rightarrow (168.03,\ 190.83)$

You can be 95% confident that the height will be between 168.03 centimeters and 190.83 centimeters when the metacarpal bone length is 50 centimeters.

## 11.1 THE SIGN TEST

## 11.1 Try It Yourself Solutions

**1a.** The claim is "the median number of days a home is on the market in its city is greater than 120."

$H_0$ : median $\leq 120$; $H_a$ : median $> 120$ (claim)

**b.** $\alpha = 0.025$

**c.** $n = 24$

**d.** The critical value is 6.

**e.** $x = 6$

**f.** Because $x \leq 6$, reject $H_0$.

**g.** There is enough evidence at the 2.5% level of significance to support the agency's claim that the median number of days a home is on the market in its city is greater than 120.

**2a.** The claim is "the median age of museum workers in the United States is 40 years old."

$H_0$ : median $= 40$ (claim); $H_a$ : median $\neq 40$

**b.** $\alpha = 0.10$

**c.** $n = 91$

**d.** The critical value is $z_0 = -1.645$.

**e.** $x = 45$

$$z = \frac{(x + 0.5) - 0.5(n)}{\frac{\sqrt{n}}{2}} = \frac{(45 + 0.5) - 0.5(91)}{\frac{\sqrt{91}}{2}} \approx \frac{0}{4.7697} \approx 0$$

**f.** Because $z > -1.645$, fail to reject $H_0$.

**g.** There is not enough evidence at the 10% level of significance to reject the organization's claim that the median age of museum workers in the United States is 40 years old.

**3a.** The claim is "a new vaccine will decrease the number of colds in adults."

$H_0$ : The number of colds will not decrease.

$H_a$ : The number of colds will decrease. (claim)

**b.** $\alpha = 0.05$

**c.** $n = 11$

**d.** The critical value is 2.

**e.** $x = 2$

**f.** Because $x \leq 2$ R, reject $H_0$.

**g.** There is enough evidence at the 5% level of significance to support the researcher's claim that the new vaccine will decrease the number of colds in adults.

## 11.1 EXERCISE SOLUTIONS

1.  A nonparametric test is a hypothesis test that does not require any specific conditions concerning the shapes of populations or the values of any population parameters.

    A nonparametric test is usually easier to perform than its corresponding parametric test, but the nonparametric test is usually less efficient.

3.  When $n$ is less than or equal to 25, the test statistic is equal to $x$ (the smaller number of $+$ or $-$ signs). When $n$ is greater than 25, the test statistic is equal to $z = \dfrac{(x+0.5)-0.5n}{\dfrac{\sqrt{n}}{2}}$.

5.  Verify that the sample is random. Identify the claim and state $H_0$ and $H_a$. Identify the level of significance and sample size. Find the critical value using Table 8 (if $n \le 25$) or Table 4 ($n > 25$). Calculate the test statistic. Make a decision and interpret it in the context of the problem.

7.  (a) The claim is "the median amount of new credit card charges for the previous month was more than \$300."

    $H_0$: median $\le$ \$300; $H_a$: median $>$ \$300 (claim)

    (b) The critical value is 1.

    (c) $x = 5$

    (d) Because $x > 1$, fail to reject $H_0$.

    (e) There is not enough evidence at the 1% level of significance for the accountant to conclude that the median amount of new credit charges for the previous month was more than \$300.

9.  (a) The claim is "the median sales price of new privately owned one-family homes sold in a recent month is \$193,000 or less."

    $H_0$: median $\le$ \$193,000 (claim); $H_a$: median $>$ \$193,000

    (b) The critical value is 1.

    (c) $x = 4$

    (d) Because $x > 1$, fail to reject $H_0$.

    (e) There is not enough evidence at the 5% level of significance to reject the agent's claim that the median sales price of new privately owned one-family homes sold in a recent month is \$193,000 or less.

11. (a) The claim is "the median amount of credit card debt for families holding such debt is at least \$2600."

    $H_0$: median $\ge$ \$2600 (claim); $H_a$: median $<$ \$2600

    (b) The critical value is $z_0 = -2.05$.

    (c) $x = 44$

    $$z = \frac{(x+0.5)-0.5(n)}{\dfrac{\sqrt{n}}{2}} = \frac{(44+0.5)-0.5(104)}{\dfrac{\sqrt{104}}{2}} \approx \frac{-7.5}{5.099} \approx -1.47$$

    (d) Because $z > -2.05$, fail to reject $H_0$.

    (e) There is not enough evidence at the 2% level of significance to reject the institution's claim that the median amount of credit card debt for families holding such debts is at least \$2600.

**13.** (a) The claim is "the median age of Twitter® users is greater than 30 years old."

$H_0$: median $\leq 30$; $H_a$: median $> 30$ (claim)

   (b) The critical value is 4.

   (c) $x = 10$

   (d) Because $x > 4$, fail to reject $\hat{H}_0$.

   (e) There is not enough evidence at the 1% level of significance to support the research group's claim that the median age of Twitter® users is greater than 30 years old.

**15.** (a) The claim is "the median number of rooms in renter-occupied units is four."

$H_0$: median $= 4$; (claim) $H_a$: median $\neq 4$

   (b) The critical value is $z_0 = -1.96$.

   (c) $x = 29$

$$z = \frac{(x + 0.5) - 0.5(n)}{\dfrac{\sqrt{n}}{2}} = \frac{(29 + 0.5) - 0.5(82)}{\dfrac{\sqrt{82}}{2}} \approx \frac{-11.5}{4.53} \approx -2.54$$

   (d) Because $z < -1.96$, reject $H_0$.

   (e) There is enough evidence at the 5% level of significance to reject the organization's claim that the median number of rooms in renter-occupied units is four.

**17.** (a) The claim is "the median hourly wage of computer systems analysts is \$38.31."

$H_0$: median $= \$38.31$; (claim) $H_a$: median $\neq \$38.31$

   (b) The critical value is $z_0 = -2.575$.

   (c) $x = 18$

$$z = \frac{(x + 0.5) - 0.5(n)}{\dfrac{\sqrt{n}}{2}} = \frac{(18 + 0.5) - 0.5(43)}{\dfrac{\sqrt{43}}{2}} \approx \frac{-3}{3.28} \approx -0.91$$

   (d) Because $z > -2.575$, fail to reject $H_0$.

   (e) There is not enough evidence at the 1% level of significance to reject the labor organization's claim that the median hourly wage of computer systems analysts is \$38.31.

**19.** (a) The claim is "the lower back pain intensity scores will decrease after acupuncture treatment."

$H_0$: The lower back pain intensity scores will not decrease.

$H_a$: The lower back pain intensity scores will decrease. (claim)

   (b) The critical value is 1.

   (c) $x = 0$

   (d) Because $x \leq 1$, reject $H_0$.

   (e) There is enough evidence at the 5% level of significance to support the physician's claim that lower back pain intensity scores will decrease after receiving acupuncture treatment.

**21.** (a) The claim is "the student's critical reading SAT scores will improve."

$H_0$: The SAT scores will not improve.

$H_a$: The SAT scores will improve. (claim)

   (b) The critical value is 1.

   (c) $x = 1$

   (d) Because $x \leq 1$, reject $H_0$.

(e) There is enough evidence at the 5% level of significance to support the agency's claim that the critical reading SAT scores will improve.

23. (a) The claim is "the proportion of adults who feel older than their real age is equal to the proportion of adults who feel younger than their real age."

$H_0$ : The proportion of adults who feel older than their real age is equal to the proportion of adults who feel younger than their real age. (claim)

$H_a$ : The proportion of adults who feel older than their real age is different from the proportion of adults who feel younger than their real age.

The critical value is 3.

$x = 3$

Because $x \le 3$, reject $H_0$.

(b) There is enough evidence at the 5% level of significance to reject the claim that the proportion of adults who feel older than their real age is equal to the proportion of adults who feel younger than their real age.

25. (a) The claim is "the median weekly earnings of female workers are less than or equal to $704."

$H_0$: median $\le$ \$704; (claim) $H_a$: median > \$704

(b) The critical value is $z_0 = 2.33$.

(c) $x = 29$

$$z = \frac{(x - 0.5) - 0.5(n)}{\frac{\sqrt{n}}{2}} = \frac{(29 - 0.5) - 0.5(47)}{\frac{\sqrt{47}}{2}} \approx \frac{5}{3.428} \approx 1.46$$

(d) Because $z < 2.33$, fail to reject $H_0$.

(e) There is not enough evidence at the 1% level of significance to reject the organization's claim that the median weekly earnings of female workers is less than or equal to $704.

27. (a) The claim is "the median age of brides at the time of their first marriage is less than or equal to 27 years old."

$H_0$: median $\le$ 27 (claim); $H_a$: median > 27

(b) The critical value is $z_0 = 1.645$.

(c) $x = 35$

$$z = \frac{(x - 0.5) - 0.5(n)}{\frac{\sqrt{n}}{2}} = \frac{(35 - 0.5) - 0.5(59)}{\frac{\sqrt{59}}{2}} \approx \frac{5}{3.841} \approx 1.30$$

(d) Because $z < 1.645$, fail to reject $H_0$.

(e) There is not enough evidence at the 5% level of significance to reject the counselor's claim that the median age of brides at the time of their first marriage is less than or equal to 27 years old.

## 11.2 THE WILCOXON TESTS

## 11.2 Try It Yourself Solutions

**1a.** The claim is "a spray-on water repellant is effective."

$H_0$: The water repellent does not increase the water repelled.

$H_a$: The water repellent increases the water repelled. (claim)

**b.** $\alpha = 0.01$

**c.** $n = 11$

**d.** The critical value is 7.

**e.**

| No repellent | Repellent applied | Difference | Absolute value | Rank | Signed rank |
|---|---|---|---|---|---|
| 8 | 15 | −7 | 7 | 11 | −11 |
| 7 | 12 | −5 | 5 | 9 | −9 |
| 7 | 11 | −4 | 4 | 7.5 | −7.5 |
| 4 | 6 | −2 | 2 | 3.5 | −3.5 |
| 6 | 6 | 0 | 0 | — | — |
| 10 | 8 | 2 | 2 | 3.5 | 3.5 |
| 9 | 8 | 1 | 1 | 1.5 | 1.5 |
| 5 | 6 | −1 | 1 | 1.5 | −1.5 |
| 9 | 12 | −3 | 3 | 5.5 | −5.5 |
| 11 | 8 | 3 | 3 | 5.5 | 5.5 |
| 8 | 14 | −6 | 6 | 10 | −10 |
| 4 | 8 | −4 | 4 | 7.5 | −7.5 |

Sum of negative ranks = −55.5

Sum of positive ranks = 10.5

$w_s = 10.5$

**f.** Because $w_s > 7$, fail to reject $H_0$.

**g.** There is not enough evidence at the 1% level of significance to support the claim that the spray-on water repellent is effective.

**2a.** The claim is "there is a difference in the claims paid by paid by the companies."

$H_0$: There is no difference in the claims paid by paid by the companies.

$H_a$: There is a difference in the claims paid by paid by the companies. (claim)

**b.** $\alpha = 0.05$

**c.** The critical values are $z_0 = \pm 1.96$.

**d.** $n_1 = 12$ and $n_2 = 12$

**e.**

| Ordered data | Sample | Rank | | Ordered data | Sample | Rank |
|---|---|---|---|---|---|---|
| 1.7 | B | 1 | | 5.3 | B | 13 |
| 1.8 | B | 2 | | 5.6 | B | 14 |
| 2.2 | B | 3 | | 5.8 | A | 15 |
| 2.5 | A | 4 | | 6.0 | A | 16 |
| 3.0 | A | 5.5 | | 6.2 | A | 17 |
| 3.0 | B | 5.5 | | 6.3 | A | 18 |
| 3.4 | B | 7 | | 6.5 | A | 19 |
| 3.9 | A | 8 | | 7.3 | B | 20 |
| 4.1 | B | 9 | | 7.4 | A | 21 |
| 4.4 | B | 10 | | 9.9 | A | 22 |
| 4.5 | A | 11 | | 10.6 | A | 23 |
| 4.7 | B | 12 | | 10.8 | B | 24 |

$R = $ sum ranks of company B $= 120.5$ *(or* $R = 179.5$*)*

**f.** $\mu_R = \dfrac{n_1(n_1 + n_2 + 1)}{2} = \dfrac{12(12 + 12 + 1)}{2} = 150$

$\sigma_R = \sqrt{\dfrac{n_1 n_2 (n_1 + n_2 + 1)}{12}} = \sqrt{\dfrac{(12)(12)(12 + 12 + 1)}{12}} \approx 17.321$

$z = \dfrac{R - \mu_R}{\sigma_R} \approx \dfrac{120.5 - 150}{17.321} \approx -1.703$ *(or* $z = 1.703$*)*

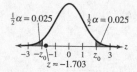

$\frac{1}{2}\alpha = 0.025 \qquad \frac{1}{2}\alpha = 0.025$

$z \approx -1.703$

**g.** Because $-1.96 < z < 1.96$, fail to reject $H_0$.

**h.** There is not enough evidence at the 5% level of significance to conclude that there is a difference in the claims paid by the companies.

## 11.2 EXERCISE SOLUTIONS

**1.** When the samples are dependent, use the Wilcoxon signed-rank test. When the samples are independent, use the Wilcoxon rank sum test.

**3.** (a) The claim is "there was no reduction in diastolic blood pressure."

   $H_0$: There is no reduction in diastolic blood pressure. (claim)

   $H_a$: There is a reduction in diastolic blood pressure.

   (b) Wilcoxon signed-rank test
   (c) The critical value is 10.

(d)

| Before treatment | After treatment | Difference | Absolute difference | Rank | Signed rank |
|---|---|---|---|---|---|
| 108 | 99 | 9 | 9 | 8 | 8 |
| 109 | 115 | −6 | 6 | 4.5 | −4.5 |
| 120 | 105 | 15 | 15 | 12 | 12 |
| 129 | 116 | 13 | 13 | 10.5 | 10.5 |
| 112 | 115 | −3 | 3 | 2 | −2 |
| 111 | 117 | −6 | 6 | 4.5 | −4.5 |
| 117 | 108 | 9 | 9 | 8 | 8 |
| 135 | 122 | 13 | 13 | 10.5 | 10.5 |
| 124 | 120 | 4 | 4 | 3 | 3 |
| 118 | 126 | −8 | 8 | 6 | −6 |
| 130 | 128 | 2 | 2 | 1 | 1 |
| 115 | 106 | 9 | 9 | 8 | 8 |

Sum of negative ranks = −17

Sum of positive ranks = 61

$w_s = 17$

(e) Because $w_s > 10$, fail to reject $H_0$.

(f) There is not enough evidence at the 1% level of significance to reject the claim that there was no reduction in diastolic blood pressure.

5. (a) The claim is "there is a difference in the earnings of people with bachelor's degrees and those with associate's degrees."

   $H_0$: There is no difference in the earnings.

   $H_a$: There is a difference in the earnings. (claim)

   (b) Wilcoxon rank sum test

   (c) The critical values are $z_0 = \pm 1.96$.

(d)

| Ordered data | Sample | Rank |
|:---:|:---:|:---:|
| 46 | B | 1 |
| 52 | B | 2.5 |
| 52 | B | 2.5 |
| 54 | B | 4 |
| 56 | B | 5.5 |
| 56 | B | 5.5 |
| 58 | B | 7 |
| 62 | B | 8 |
| 65 | B | 9 |
| 72 | B | 10 |
| 78 | B | 11 |
| 81 | A | 12 |
| 82 | A | 13 |
| 84 | A | 14 |
| 86 | A | 15.5 |
| 86 | A | 15.5 |
| 87 | A | 17 |
| 90 | A | 18 |
| 93 | A | 19.5 |
| 93 | A | 19.5 |
| 95 | A | 21 |

$R$ = sum ranks of bachelor's degree = 165

$$\mu_R = \frac{n_1(n_1 + n_2 + 1)}{2} = \frac{10(10 + 11 + 1)}{2} = 110$$

$$\sigma_R = \sqrt{\frac{n_1 n_2 (n_1 + n_2 + 1)}{12}} = \sqrt{\frac{(10)(11)(10 + 11 + 1)}{12}} \approx 14.201$$

$$z = \frac{R - \mu_R}{\sigma_R} \approx \frac{165 - 110}{14.201} \approx 3.87$$

(e) Because $z > 1.96$, reject $H_0$.

(f) There is enough evidence at the 5% level of significance to support the administrator's claim that there is a difference in the earnings of people with bachelor's degrees and those with advanced degrees.

7. (a) The claim is "there is a difference in the salaries earned by teachers in Wisconsin and Michigan."

$H_0$: There is not a difference in the salaries.

$H_a$: There is a difference in the salaries. (claim)

(b) Wilcoxon rank sum test

(c) The critical values are $z_0 = \pm 1.96$.

(d)

| Ordered data | Sample | Rank |
|:---:|:---:|:---:|
| 47 | WI | 1 |
| 49 | WI | 2 |
| 51 | WI | 3 |
| 52 | WI | 4 |
| 53 | WI | 5 |
| 55 | WI | 6.5 |
| 55 | WI | 6.5 |
| 56 | MI | 8.5 |
| 56 | WI | 8.5 |
| 58 | MI | 10 |
| 59 | WI | 11 |
| 60 | MI | 12 |
| 61 | WI | 14 |
| 61 | MI | 14 |
| 61 | WI | 14 |
| 62 | MI | 16 |
| 64 | MI | 17.5 |
| 64 | MI | 17.5 |
| 65 | MI | 19 |
| 68 | MI | 20 |
| 70 | MI | 21.5 |
| 70 | MI | 21.5 |
| 79 | MI | 23 |

$R$ = sum ranks of Wisconsin = 75.5

$$\mu_R = \frac{n_1(n_1+n_2+1)}{2} = \frac{11(11+12+1)}{2} = 132$$

$$\sigma_R = \sqrt{\frac{n_1 n_2 (n_1 + n_2 + 1)}{12}} = \sqrt{\frac{(11)(12)(11+12+1)}{12}} \approx 16.248$$

$$z = \frac{R - \mu_R}{\sigma_R} \approx \frac{75.5 - 132}{16.248} \approx -3.48$$

(e) Because $z < -1.96$, reject $H_0$.

(f) There is enough evidence at the 5% level of significance to support the representative's claim that there is a difference in the salaries earned by teachers in Wisconsin and Michigan.

**9.** The claim is "a certain fuel additive improves a car's gas mileage."

$H_0$: The fuel additive does not improve gas mileage.

$H_a$: The fuel additive does improve gas mileage. (claim)

The critical value is $z_0 = -1.28$.

| Before | After | Difference | Absolute value | Rank | Signed rank |
|--------|-------|------------|----------------|------|-------------|
| 36.4 | 36.7 | −0.3 | 0.3 | 4.5 | −4.5 |
| 36.4 | 36.9 | −0.5 | 0.5 | 11 | −11 |
| 36.6 | 37.0 | −0.4 | 0.4 | 7 | −7 |
| 36.6 | 37.5 | −0.9 | 0.9 | 17 | −17 |
| 36.8 | 38.0 | −1.2 | 1.2 | 19.5 | −19.5 |
| 36.9 | 38.1 | −1.2 | 1.2 | 19.5 | −19.5 |
| 37.0 | 38.4 | −1.4 | 1.4 | 25 | −25 |
| 37.1 | 38.7 | −1.6 | 1.6 | 30.5 | −30.5 |
| 37.2 | 38.8 | −1.6 | 1.6 | 30.5 | −30.5 |
| 37.2 | 38.9 | −1.7 | 1.7 | 32 | −32 |
| 36.7 | 36.3 | 0.4 | 0.4 | 7 | 7 |
| 37.5 | 38.9 | −1.4 | 1.4 | 25 | −25 |
| 37.6 | 39.0 | −1.4 | 1.4 | 25 | −25 |
| 37.8 | 39.1 | −1.3 | 1.3 | 21.5 | −21.5 |
| 37.9 | 39.4 | −1.5 | 1.5 | 28.5 | −28.5 |
| 37.9 | 39.4 | −1.5 | 1.5 | 28.5 | −28.5 |
| 38.1 | 39.5 | −1.4 | 1.4 | 25 | −25 |
| 38.4 | 39.8 | −1.4 | 1.4 | 25 | −25 |
| 40.2 | 40.0 | 0.2 | 0.2 | 2.5 | 2.5 |
| 40.5 | 40.0 | 0.5 | 0.5 | 11 | 11 |
| 40.9 | 40.1 | 0.8 | 0.8 | 16 | 16 |
| 35.0 | 36.3 | −1.3 | 1.3 | 21.5 | −21.5 |
| 32.7 | 32.8 | −0.1 | 0.1 | 1 | −1 |
| 33.6 | 34.2 | −0.6 | 0.6 | 14.5 | −14.5 |
| 34.2 | 34.7 | −0.5 | 0.5 | 11 | −11 |
| 35.1 | 34.9 | 0.2 | 0.2 | 2.5 | 2.5 |
| 35.2 | 34.9 | 0.3 | 0.3 | 4.5 | 4.5 |
| 35.3 | 35.3 | 0 | 0 | — | — |
| 35.5 | 35.9 | −0.4 | 0.4 | 7 | −7 |
| 35.9 | 36.4 | −0.5 | 0.5 | 11 | −11 |
| 36.0 | 36.6 | −0.6 | 0.6 | 14.5 | −14.5 |
| 36.1 | 36.6 | −0.5 | 0.5 | 11 | −11 |
| 37.2 | 38.3 | −1.1 | 1.1 | 18 | −18 |

Sum of negative ranks $= -484.5$

Sum of positive ranks $= 43.5$

$w_s = 43.5$

$$z = \frac{w_s - \dfrac{n(n+1)}{4}}{\sqrt{\dfrac{n(n+1)(2n+1)}{24}}} = \frac{43.5 - \dfrac{32(32+1)}{4}}{\sqrt{\dfrac{32(32+1)\left[(2)32+1\right]}{24}}} = \frac{-220.5}{\sqrt{2860}} = -4.123$$

Note: $n = 32$ because one of the differences is zero and should be discarded.

Because $z < -1.28$, reject $H_0$. There is enough evidence at the 10% level of significance for the engineer to conclude that the gas mileage is improved.

## 11.3 THE KRUSKAL-WALLIS TEST

## 11.3 Try It Yourself Solutions

**1a.** The claim is "the distribution of the veterinarians' salaries in at least one of the three states is different from the others."

$H_0$: There is no difference in the salaries in the three states.

$H_a$: There is a difference in the salaries in the three states. (claim)

**b.** $\alpha = 0.05$

**c.** d.f. $= k - 1 = 2$

**d.** $\chi_0^2 = 5.991$; Rejection region: $\chi^2 > 5.991$

**e.**

| Ordered data | State | Rank | | Ordered data | State | Rank |
|---|---|---|---|---|---|---|
| 84.5 | OH | 1 | | 98.5 | TX | 15 |
| 87.9 | TX | 2 | | 98.7 | FL | 16 |
| 90.9 | OH | 3 | | 99.4 | TX | 17.5 |
| 93.0 | OH | 4 | | 99.4 | OH | 17.5 |
| 93.2 | OH | 5 | | 99.6 | TX | 19 |
| 93.3 | FL | 6 | | 100.4 | TX | 20 |
| 93.9 | FL | 7 | | 100.6 | FL | 21 |
| 94.9 | OH | 8 | | 100.9 | TX | 22 |
| 95.2 | FL | 9 | | 102.4 | FL | 23 |
| 95.7 | OH | 10 | | 102.8 | FL | 24 |
| 95.9 | TX | 11 | | 102.9 | TX | 25 |
| 96.3 | OH | 12 | | 103.2 | FL | 26 |
| 97.2 | TX | 13 | | 106.2 | OH | 27 |
| 98.3 | FL | 14 | | 113.6 | TX | 28 |

$R_1 = 172.5$, $R_2 = 146$, $R_3 = 87.5$

**f.** $H = \dfrac{12}{N(N+1)} \left( \dfrac{R_1^2}{n_1} + \dfrac{R_2^2}{n_2} + \dfrac{R_3^2}{n_3} \right) - 3(N+1)$

$= \dfrac{12}{28(29)} \left( \dfrac{(172.5)^2}{10} + \dfrac{(146)^2}{9} + \dfrac{(87.5)^2}{9} \right) - 3(29)$

$\approx 4.548$

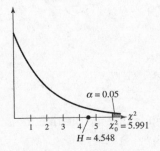

$\alpha = 0.05$

$\chi_0^2 = 5.991$

$H \approx 4.548$

**g.** Because $H < 5.991$, fail to reject $H_0$.

**h.** There is not enough evidence at the 5% level of significance to conclude that the distribution of the salaries in at least one state is different from the others.

## 11.3 EXERCISE SOLUTIONS

1. The conditions for using a Kruskal-Wallis test are that the samples must be random and independent, and the size of each sample must be at least 5.

3. (a) The claim is "the distributions of the annual premiums in at least one state is different from the others."

$H_0$: The distribution of the annual premiums is the same in all three states.

$H_a$: The distribution of the annual premiums in at least one state is different from the others. (claim)

(b) $\chi_0^2 = 5.991$; Rejection region: $\chi^2 > 5.991$

(c)

| Ordered data | Sample | Rank |
|---|---|---|
| 605 | VA | 1 |
| 616 | VA | 2 |
| 688 | VA | 3 |
| 695 | VA | 4 |
| 784 | MA | 5 |
| 800 | VA | 6 |
| 848 | CT | 7 |
| 885 | VA | 8 |
| 916 | MA | 9 |
| 929 | CT | 10 |
| 982 | VA | 11 |
| 1007 | MA | 12 |
| 1013 | CT | 13 |
| 1052 | MA | 14 |
| 1053 | CT | 15 |
| 1070 | CT | 16 |
| 1132 | MA | 17 |
| 1137 | MA | 18 |
| 1163 | CT | 19 |
| 1288 | CT | 20 |
| 1322 | MA | 21 |

$R_1 = 100,\ R_2 = 96,\ R_3 = 35$

$$H = \frac{12}{N(N+1)} \left( \frac{R_1^2}{n_1} + \frac{R_2^2}{n_2} + \frac{R_3^2}{n_3} \right) - 3(N+1)$$

$$= \frac{12}{21(21+1)} \left( \frac{(100)^2}{7} + \frac{(96)^2}{7} + \frac{(35)^2}{7} \right) - 3(21+1)$$

$$\approx 9.848$$

(d) Because $H > 5.991$, reject $H_0$.

(e) There is enough evidence at the 5% level of significance to conclude that the distribution of the annual premiums in at least one state is different from the others.

**5.** (a) The claim is "the distribution of the annual salaries of private industry workers in at least one state is different from the others."

$H_0$: The distribution of the annual salaries is the same in all four states.

$H_a$: The distribution of the annual salaries in at least one state is different from the others. (claim)

(b) $\chi_0^2 = 6.251$; Rejection region: $\chi^2 > 6.251$

(c)

| Ordered data | Sample | Rank |
|:---:|:---:|:---:|
| 28.3 | KY | 1 |
| 29.8 | SC | 2.5 |
| 29.8 | SC | 2.5 |
| 31.6 | WV | 4.5 |
| 31.6 | WV | 4.5 |
| 33.4 | WV | 6 |
| 33.7 | KY | 7 |
| 34.7 | SC | 8 |
| 34.9 | WV | 9 |
| 35.3 | KY | 10.5 |
| 35.3 | KY | 10.5 |
| 35.5 | NC | 12 |
| 36.1 | SC | 13 |
| 36.6 | NC | 14 |
| 37.0 | KY | 15 |
| 37.4 | SC | 16 |
| 39.6 | NC | 17 |
| 41.9 | NC | 18.5 |
| 41.9 | WV | 18.5 |
| 42.7 | WV | 20 |
| 42.9 | SC | 21 |
| 43.5 | NC | 23 |
| 43.5 | NC | 23 |
| 43.5 | SC | 23 |
| 45.9 | KY | 25 |
| 47.1 | WV | 26 |
| 54.3 | NC | 27 |
| 57.5 | KY | 28 |

$R_1 = 97$, $R_2 = 134.5$, $R_3 = 86$, $R_4 = 88.5$

$$H = \frac{12}{N(N+1)}\left(\frac{R_1^2}{n_1} + \frac{R_2^2}{n_2} + \frac{R_3^2}{n_3} + \frac{R_4^2}{n_4}\right) - 3(N+1)$$

$$= \frac{12}{28(28+1)}\left(\frac{(97)^2}{7} + \frac{(134.5)^2}{7} + \frac{(86)^2}{7} + \frac{(88.5)^2}{7}\right) - 3(28+1)$$

$$\approx 3.206$$

(d) Because $H < 6.251$, fail to reject $H_0$.

(e) There is not enough evidence at the 10% level of significance to conclude that the distribution of the annual salaries in at least one state is different from the others.

**7. (a)** The claim is "the number of days patients spend in the hospital is different in at least one region of the United States."

$H_0$: The number of days spent in the hospital is the same for all four regions.

$H_a$: The number of days spent in the hospital is different in at least one region. (claim)

$\chi_0^2 = 11.345$; Rejection region: $\chi^2 > 11.345$

| Ordered data | Sample | Rank | | Ordered data | Sample | Rank |
|---|---|---|---|---|---|---|
| 1 | NE | 2.5 | | 5 | S | 19 |
| 1 | MW | 2.5 | | 5 | W | 19 |
| 1 | S | 2.5 | | 5 | W | 19 |
| 1 | S | 2.5 | | 6 | NE | 26 |
| 2 | W | 5 | | 6 | NE | 26 |
| 3 | NE | 8.5 | | 6 | NE | 26 |
| 3 | NE | 8.5 | | 6 | MW | 26 |
| 3 | MW | 8.5 | | 6 | W | 26 |
| 3 | MW | 8.5 | | 6 | W | 26 |
| 3 | W | 8.5 | | 6 | W | 26 |
| 3 | W | 8.5 | | 7 | MW | 30.5 |
| 4 | MW | 13.5 | | 7 | S | 30.5 |
| 4 | MW | 13.5 | | 8 | NE | 33.5 |
| 4 | MW | 13.5 | | 8 | NE | 33.5 |
| 4 | W | 13.5 | | 8 | S | 33.5 |
| 5 | NE | 19 | | 8 | S | 33.5 |
| 5 | MW | 19 | | 9 | MW | 36 |
| 5 | S | 19 | | 11 | NE | 37 |
| 5 | S | 19 | | | | |

$R_1 = 220.5$, $R_2 = 171.5$, $R_3 = 159.5$, $R_4 = 151.5$

$$H = \frac{12}{N(N+1)}\left(\frac{R_1^2}{n_1} + \frac{R_2^2}{n_2} + \frac{R_3^2}{n_3} + \frac{R_4^2}{n_4}\right) - 3(N+1)$$

$$= \frac{12}{37(37+1)}\left(\frac{(220.5)^2}{10} + \frac{(171.5)^2}{10} + \frac{(159.5)^2}{8} + \frac{(151.5)^2}{9}\right) - 3(37+1)$$

$$\approx 1.507$$

Because $H < 11.345$, fail to reject $H_0$. There is not enough evidence at the 1% level of significance to support the underwriter's claim that the number of days patients spend in the hospital is different in at least one region of the United States.

**(b)**

| Variation | Sum of squares | Degrees of freedom | Mean squares | F | p |
|---|---|---|---|---|---|
| Between | 9.17 | 3 | 3.06 | 0.52 | 0.673 |
| Within | 194.72 | 33 | 5.90 | | |

Because $\alpha = 0.01$, the critical value is about 4.51. Because $F \approx 0.52$ is less than the critical value, the decision is to fail to reject $H_0$. There is not enough evidence at the 1% level of significance to support the underwriter's claim that the number of days patients spend in the hospital is different in at least one region of the United States.

(c) Both tests come to the same decision, which is that there is not enough evidence to support the claim that the number of days patients spend in the hospital is different in at least one region of the United States.

## 11.4 RANK CORRELATION

## 11.4 Try It Yourself Solutions

**1a.** The claim is "there is a significant correlation between the oat and wheat prices."

$H_0: \rho_s = 0$ ; $H_a$: $\rho_s \neq 0$ (claim)

**b.** $\alpha = 0.10$

**c.** The critical value is 0.714.

**d.**

| Oat | Rank | Wheat | Rank | $d$ | $d^2$ |
|-----|------|-------|------|-----|-------|
| 4.04 | 3.5 | 7.96 | 3 | -0.5 | 0.25 |
| 4.38 | 7 | 8.13 | 7 | 0 | 0 |
| 4.03 | 2 | 7.72 | 1 | 1 | 1 |
| 4.05 | 5 | 7.97 | 4 | 1 | 1 |
| 4.21 | 6 | 8.01 | 6 | 0 | 0 |
| 4.02 | 1 | 7.75 | 2 | -1 | 1 |
| 4.04 | 3.5 | 7.98 | 5 | -1.5 | 2.25 |
| | | | | | $\sum d^2 = 5.5$ |

**e.** $r_s \approx 1 - \dfrac{6\sum d^2}{n(n^2 - 1)} = 1 - \dfrac{6(5.5)}{7(7^2 - 1)} \approx 0.902$

**f.** Because $r_s > 0.714$ , reject $H_0$.

**g.** There is enough evidence at the 10% level of significance to conclude that there is a significant correlation between the oat and wheat prices.

## 11.4 EXERCISE SOLUTIONS

1. The Spearman rank correlation coefficient can be used to describe the relationship between linear and nonlinear data. Also, it can be used for data at the ordinal level and it is easier to calculate by hand than the Pearson correlation coefficient.

3. The ranks of corresponding data pairs are exactly identical when $r_s$ is equal to 1.

    The ranks are in "reverse" order when $r_s$ is equal to –1.

    The ranks of have no relationship when $r_s$ is equal to 0.

5. (a) The claim is "there is a significant correlation between purchased seed expenses and fertilizer and lime expenses in the farming business."

    $H_0: \rho_s = 0$ ; $H_a: \rho_s \neq 0$ (claim)

    (b) The critical value is 0.738.

(c)

| Seed expenses | Rank | Fertilizer/lime expense | Rank | $d$ | $d^2$ |
|---|---|---|---|---|---|
| 430 | 5 | 490 | 4 | 1 | 1 |
| 1070 | 8 | 1640 | 8 | 0 | 0 |
| 330 | 3 | 520 | 5 | -2 | 4 |
| 164 | 1 | 360 | 1 | 0 | 0 |
| 610 | 6 | 557 | 6 | 0 | 0 |
| 340 | 4 | 460 | 3 | 1 | 1 |
| 710 | 7 | 893 | 7 | 0 | 0 |
| 250 | 2 | 380 | 2 | 0 | 0 |
| | | | | | $\sum d^2 = 6$ |

$$r_s \approx 1 - \frac{6\sum d^2}{n(n^2-1)} \approx 1 - \frac{6(6)}{8(8^2-1)} \approx 0.929$$

(d) Because $r_s > 0.738$, reject $H_0$.

(e) There is enough evidence at the 1% level of significance to conclude that there is a significant correlation between purchased seed expenses and fertilizer and lime expenses.

7. (a) The claim is "there is a significant correlation between the barley and corn prices."
   $H_0: \rho_s = 0$; $H_a: \rho_s \neq 0$ (claim)

   (b) The critical value is 0.700.

   (c)

| Barley | Rank | Corn | Rank | $d$ | $d^2$ |
|---|---|---|---|---|---|
| 5.42 | 3 | 6.05 | 1 | 2 | 4 |
| 5.40 | 2 | 6.28 | 2 | 0 | 0 |
| 5.35 | 1 | 6.34 | 3 | -2 | 4 |
| 5.70 | 5 | 6.36 | 5.5 | -0.5 | 0.25 |
| 5.72 | 6 | 6.36 | 5.5 | 0.5 | 0.25 |
| 5.48 | 4 | 6.35 | 4 | 0 | 0 |
| 6.33 | 7 | 7.16 | 8 | -1 | 1 |
| 6.45 | 8 | 7.65 | 9 | -1 | 1 |
| 6.46 | 9 | 6.90 | 7 | 2 | 4 |
| | | | | | $\sum d^2 = 14.5$ |

$$r_s \approx 1 - \frac{6\sum d^2}{n(n^2-1)} \approx 1 - \frac{6(14.5)}{9(9^2-1)} \approx 0.879$$

(d) Because $r_s > 0.700$, reject $H_0$.

(e) There is enough evidence at the 5% level of significance to conclude that there is a significant correlation between the barley and corn prices.

9. The claim is "there is a significant correlation between science achievement scores and GNI."
   $H_0: \rho_s = 0$; $H_a: \rho_s \neq 0$ (claim)
   The critical value is 0.600.

| Science average | Rank | GNI | Rank | $d$ | $d^2$ |
|---|---|---|---|---|---|
| 527 | 8 | 1317 | 3 | 5 | 25 |
| 497 | 6 | 2668 | 6 | 0 | 0 |
| 513 | 7 | 3379 | 7 | 0 | 0 |
| 483 | 2.5 | 2101 | 5 | -2.5 | 6.25 |
| 529 | 9 | 5170 | 8 | 1 | 1 |
| 419 | 1 | 866 | 2 | -1 | 1 |
| 483 | 2.5 | 1429 | 4 | -1.5 | 2.25 |
| 494 | 5 | 414 | 1 | 4 | 16 |
| 487 | 4 | 13,924 | 9 | -5 | 25 |
| | | | | | $\sum d^2 = 76.5$ |

$$r_s \approx 1 - \frac{6\sum d^2}{n(n^2-1)} \approx 1 - \frac{6(76.5)}{9(9^2-1)} \approx 0.363$$

Because $r_s < 0.600$, fail to reject $H_0$. There is not enough evidence at the 10% level of significance to conclude that there is a significant correlation between science achievement scores and GNI.

11. The claim is "there is a significant correlation between science and mathematics achievement scores."

$H_0: \rho_s = 0$; $H_a: \rho_s \neq 0$ (claim)

The critical value is 0.600.

| Science average | Rank | Math average | Rank | $d$ | $d^2$ |
|---|---|---|---|---|---|
| 527 | 8 | 529 | 8 | 0 | 0 |
| 497 | 6 | 498 | 5 | 1 | 1 |
| 513 | 7 | 520 | 7 | 0 | 0 |
| 483 | 2.5 | 489 | 3 | -0.5 | 0.25 |
| 529 | 9 | 539 | 9 | 0 | 0 |
| 419 | 1 | 416 | 1 | 0 | 0 |
| 483 | 2.5 | 488 | 2 | 0.5 | 0.25 |
| 494 | 5 | 495 | 4 | 1 | 1 |
| 487 | 4 | 502 | 6 | -2 | 4 |
| | | | | | $\sum d^2 = 6.5$ |

$$r_s \approx 1 - \frac{6\sum d^2}{n(n^2-1)} \approx 1 - \frac{6(6.5)}{9(9^2-1)} \approx 0.946$$

Because $r_s > 0.600$, reject $H_0$. There is enough evidence at the 10% level of significance to conclude that there is a significant correlation between science and mathematics achievement scores.

13. The claim is "there is a significant correlation between average hours worked and the number of on-the-job injuries."

$H_0: \rho_s = 0$; $H_a: \rho_s \neq 0$ (claim)

The critical values are $\dfrac{\pm z}{\sqrt{n-1}} = \dfrac{\pm 1.645}{\sqrt{33-1}} \approx \pm 0.291$.

| Hours worked | Rank | Injuries | Rank | $D$ | $D^2$ |
|---|---|---|---|---|---|
| 46 | 26.5 | 22 | 14 | 12.5 | 156.25 |
| 43 | 18 | 25 | 19.5 | -1.5 | 2.25 |
| 41 | 10 | 18 | 8 | 2 | 4 |
| 40 | 5.5 | 17 | 5.5 | 0 | 0 |
| 41 | 10 | 20 | 11 | -1 | 1 |
| 42 | 15.5 | 22 | 14 | 1.5 | 2.25 |
| 45 | 22.5 | 28 | 24 | -1.5 | 2.25 |
| 45 | 22.5 | 29 | 27 | -4.5 | 20.25 |
| 42 | 15.5 | 24 | 18 | -2.5 | 6.25 |
| 45 | 22.5 | 26 | 21.5 | 1 | 1 |
| 44 | 19.5 | 26 | 21.5 | -2 | 4 |
| 44 | 19.5 | 25 | 19.5 | 0 | 0 |
| 45 | 22.5 | 27 | 23 | -0.5 | 0.25 |
| 46 | 26.5 | 29 | 27 | -0.5 | 0.25 |
| 47 | 29.5 | 29 | 27 | 2.5 | 6.25 |
| 47 | 29.5 | 30 | 31.5 | -2 | 4 |
| 46 | 26.5 | 29 | 27 | -0.5 | 0.25 |
| 46 | 26.5 | 29 | 27 | -0.5 | 0.25 |
| 49 | 31 | 30 | 31.5 | -0.5 | 0.25 |
| 50 | 32.5 | 30 | 31.5 | 1 | 1 |
| 50 | 32.5 | 30 | 31.5 | 1 | 1 |
| 42 | 15.5 | 23 | 16.5 | -1 | 1 |
| 41 | 10 | 22 | 14 | -4 | 16 |
| 42 | 15.5 | 23 | 16.5 | -1 | 1 |
| 41 | 10 | 21 | 12 | -2 | 4 |
| 41 | 10 | 19 | 10 | 0 | 0 |
| 41 | 10 | 18 | 8 | 2 | 4 |
| 41 | 10 | 18 | 8 | 2 | 4 |
| 40 | 5.5 | 17 | 5.5 | 0 | 0 |
| 39 | 3 | 16 | 2.5 | 0.5 | 0.25 |
| 38 | 1 | 16 | 2.5 | -1.5 | 2.25 |
| 39 | 3 | 16 | 2.5 | 0.5 | 0.25 |
| 39 | 3 | 16 | 2.5 | 0.5 | 0.25 |
| | | | | | $\sum d^2 = 246$ |

$$r_s \approx 1 - \frac{6\sum d^2}{n(n^2-1)} \approx 1 - \frac{6(246)}{33(33^2-1)} \approx 0.959$$

Because $r_s > 0.291$, reject $H_0$. There is enough evidence at the 10% level of significance to conclude that there is a significant correlation between average hours worked and the number of on-the-job injuries.

## 11.5 THE RUNS TEST

## 11.5 Try It Yourself Solutions

**1a.** *PPP F P F PPPP FF P F PP FFF PPP F PPP*

**b.** 13 groups $\Rightarrow$ 13 runs

**c.** 3, 1, 1, 1, 4, 2, 1, 1, 2, 3, 3, 1, 3

**2a.** The claim is "the sequence of genders is not random."

$H_0$: The sequence of genders is random.

$H_a$: The sequence of genders is not random. (claim)

**b.** $\alpha = 0.05$

**c.** *M FFF MM FF M F MM FFF*

$n_1$ = number of *F*'s = 9

$n_2$ = number of *M*'s = 6

$G$ = number of runs = 8

**d.** lower critical value = 4

upper critical value = 13

**e.** $G = 8$

**f.** Because $4 < G < 13$, fail to reject $H_0$.

**g.** There is not enough evidence at the 5% level of significance to support the claim that the sequence of genders is not random.

**3a.** The claim is "the sequence of weather conditions is not random."

$H_0$: The sequence of weather conditions is random.

$H_a$: The sequence of weather conditions is not random. (claim)

**b.** $\alpha = 0.05$

**c.** $n_1$ = number of *N*'s = 21

$n_2$ = number of *S*'s = 10

$G$ = number of runs = 17

**d.** The critical values are $z_0 = \pm 1.96$.

**e.** $\mu_G = \dfrac{2n_1 n_2}{n_1 + n_2} + 1 = \dfrac{2(21)(10)}{21 + 10} + 1 = 14.55$

$\sigma_G = \sqrt{\dfrac{2n_1 n_2 (2n_1 n_2 - n_1 - n_2)}{(n_1 + n_2)^2 (n_1 + n_2 - 1)}} = \sqrt{\dfrac{2(21)(10)(2(21)(10) - 21 - 10)}{(21 + 10)^2 (21 + 10 - 1)}} \approx 2.38$

$z = \dfrac{G - \mu_G}{\sigma_G} \approx \dfrac{17 - 14.55}{2.38} \approx 1.03$

**f.** Because $z < 1.96$, fail to reject $H_0$.

**g.** There is not enough evidence at the 5% level of significance to support the claim that the sequence of weather conditions is not random.

## 11.5 EXERCISE SOLUTIONS

**1.** Answers will vary. *Sample answer:* It is called the runs test because it considers the number of runs of data in a sample to determine whether the sequence of data was randomly selected.

**3.** Number of runs = 8

Run lengths = 1, 1, 1, 1, 3, 3, 1, 1

**5.** Number of runs = 9

Run lengths = 1, 1, 1, 1, 1, 6, 3, 2, 4

7. $n_1$ = number of $T$'s = 6

   $n_2$ = number of $F$'s = 6

9. $n_1$ = number of $M$'s = 10

   $n_2$ = number of $F$'s = 10

11. $n_1$ = number of $T$'s = 6

    $n_1$ = number of $F$'s = 6

    too high: 11; too low: 3

13. $n_1$ = number of $N$'s = 11

    $n_1$ = number of $S$'s = 7

    too high: 14; too low: 5

15. (a) The claim is "the tosses were not random."

    $H_0$: The coin tosses were random.

    $H_a$: The coin tosses were not random. (claim)

    (b) $n_1$ = number of $H$'s = 7

    $n_2$ = number of $T$'s = 9

    lower critical value = 4

    upper critical value = 14

    (c) $G = 9$ runs

    (d) Because $4 < G < 14$, fail to reject $H_0$.

    (e) There is not enough evidence at the 5% level of significance to support the claim that the coin tosses were not random.

17. (a) The claim is "the sequence of leagues of World Series winning teams is not random."

    $H_0$: The sequence of leagues of winning teams is random.

    $H_a$: The sequence of leagues of winning teams is not random. (claim)

    (b) $n_1$ = number of $N$'s = 20

    $n_2$ = number of $A$'s = 23

    The critical values are $z_0 = \pm 1.96$.

    (c) $G = 27$ runs

    $$\mu_G = \frac{2n_1 n_2}{n_1 + n_2} + 1 = \frac{2(20)(23)}{20 + 23} + 1 = 22.40$$

    $$\sigma_G = \sqrt{\frac{2n_1 n_2 (2n_1 n_2 - n_1 - n_2)}{(n_1 + n_2)^2 (n_1 + n_2 - 1)}} = \sqrt{\frac{2(20)(23)(2(20)(23) - 20 - 23)}{(20 + 23)^2 (20 + 23 - 1)}} \approx 3.22$$

    $$z = \frac{G - \mu_G}{\sigma_G} \approx \frac{27 - 22.40}{3.22} \approx 1.43$$

    (d) Because $z < 1.96$, fail to reject $H_0$.

    (e) There is not enough evidence at the 5% level of significance to conclude that the sequence of leagues of World Series winning teams is not random.

19. (a) The claim is "the microchips are random by gender."

    $H_0$: The microchips are random by gender.

    $H_a$: The microchips are not random by gender. (claim)

    (b) $n_1$ = number of $M$'s = 9

    $n_2$ = number of $F$'s = 20

    lower critical value = 8; upper critical value = 18

    (c) $G = 12$ runs

    (d) Because $8 < G < 18$, fail to reject $H_0$.

(e) There is not enough evidence at the 5% level of significance to reject the claim that the microchips are random by gender.

**21.** The claim is "the daily high temperatures do not occur randomly."

$H_0$: Daily high temperatures occur randomly.

$H_a$: Daily high temperatures do not occur randomly. (claim)

median = 87

$n_1$ = number of above = 15

$n_2$ = number of below = 13

lower critical value = 9

upper critical value = 21

$G = 11$ runs

Because $9 < G < 21$, fail to reject $H_0$,

There is not enough evidence at the 5% level of significance to support the claim that the daily high temperatures do not occur randomly.

**23.** Answers will vary.

## CHAPTER 11 REVIEW EXERCISE SOLUTIONS

**1.** (a) The claim is "the median number of customers per day is no more than 650."

$H_0$: median $\leq 650$ (claim);     $H_a$: median $> 650$

(b) The critical value is 2.

(c) $x = 7$

(d) Because $x > 2$, fail to reject $H_0$.

(e) There is not enough evidence at the 1% level of significance to reject the bank manager's claim that the median number of customers per day is no more than 650.

**3.** (a) The claim is "median sentence length for all federal prisoners is 2 years."

$H_0$: median $= 2$ (claim);     $H_a$: median $\neq 2$

(b) The critical value is $z_0 = -1.645$.

(c) $x = 65$

$$z = \frac{(x+0.5) - 0.5(n)}{\frac{\sqrt{n}}{2}} = \frac{(65+0.5) - 0.5(174)}{\frac{\sqrt{174}}{2}} \approx \frac{-21.5}{6.595} \approx -3.26$$

(d) Because $z_0 < -1.645$, reject $H_0$.

(e) There is enough evidence at the 10% level of significance to reject the agency's claim that the median sentence length for all federal prisoner is 2 years.

**5.** (a) The claim is "there was no reduction in diastolic blood pressure."

$H_0$: There is no reduction in diastolic blood pressure. (claim)

$H_a$: There is a reduction in diastolic blood pressure.

(b) The critical value is 2.

(c) $x = 3$

(d) Because $x > 2$, fail to reject $H_0$.

(e) There is not enough evidence at the 5% level of significance to reject the claim that there was no reduction in diastolic blood pressure.

7. (a) The claim is "there is a difference in the total times required to earn a doctorate degree by female and male graduate students."

   $H_0$: There is no difference in the total times to earn a doctorate degree by female and male graduate students.

   $H_a$: There is a difference in the total times to earn a doctorate degree by female and male graduate students. (claim)

   (b) Wilcoxon rank sum test

   (c) The critical values are $z_0 \pm 2.575$.

   (d)

| Ordered data | Sample | Rank | | Ordered data | Sample | Rank |
|---|---|---|---|---|---|---|
| 6 | F | 1.5 | | 9 | M | 13 |
| 6 | F | 1.5 | | 9 | M | 13 |
| 7 | M | 4.5 | | 9 | M | 13 |
| 7 | M | 4.5 | | 10 | F | 17.5 |
| 7 | M | 4.5 | | 10 | M | 17.5 |
| 7 | M | 4.5 | | 10 | M | 17.5 |
| 8 | F | 8.5 | | 10 | M | 17.5 |
| 8 | F | 8.5 | | 11 | F | 20.5 |
| 8 | M | 8.5 | | 11 | F | 20.5 |
| 8 | M | 8.5 | | 12 | F | 22.5 |
| 9 | F | 13 | | 12 | F | 22.5 |
| 9 | F | 13 | | 13 | F | 24 |

$$\mu_R = \frac{n_1(n_1 + n_2 + 1)}{2} = \frac{12(12 + 12 + 1)}{2} = 150$$

$$\sigma_R = \sqrt{\frac{n_1 n_2 (n_1 + n_2 + 1)}{12}} \sqrt{\frac{(12)(12)(12 + 12 + 1)}{12}} \approx 17.321$$

$$z = \frac{R - \mu_R}{\sigma_R} \approx \frac{126.5 - 150}{17.321} \approx -1.357$$

   (e) Because $z > -2.575$, fail to reject $H_0$.

   (f) There is not enough evidence at the 1% level of significance to support the claim that there is a difference in the total times to earn a doctorate degree by female and male graduate students.

9. (a) The claim is "the distributions of the ages of the doctorate recipients in at least one field of study is different from the others."

   $H_0$: The distribution of the ages of doctorate recipients is the same in all three fields of study.

   $H_a$: The distribution of the ages of doctorate recipients in at least one field of study is different from the others. (claim)

   (b) $\chi_0^2 = 9.210$; Rejection region: $\chi^2 > 9.210$.

(c)

| Ordered data | Sample | Rank | | Ordered data | Sample | Rank |
|---|---|---|---|---|---|---|
| 29 | L | 1.5 | | 32 | L | 20 |
| 29 | P | 1.5 | | 32 | L | 20 |
| 30 | L | 5.5 | | 32 | P | 20 |
| 30 | P | 5.5 | | 32 | P | 20 |
| 30 | P | 5.5 | | 32 | S | 20 |
| 30 | P | 5.5 | | 32 | S | 20 |
| 30 | P | 5.5 | | 33 | P | 25 |
| 30 | S | 5.5 | | 33 | S | 25 |
| 31 | L | 12.5 | | 33 | S | 25 |
| 31 | L | 12.5 | | 34 | L | 28 |
| 31 | L | 12.5 | | 34 | L | 28 |
| 31 | P | 12.5 | | 34 | S | 28 |
| 31 | P | 12.5 | | 35 | L | 31 |
| 31 | P | 12.5 | | 35 | S | 31 |
| 31 | S | 12.5 | | 35 | S | 31 |
| 31 | S | 12.5 | | 36 | S | 33 |
| 32 | L | 20 | | | | |

$R_1 = 191.5$, $R_2 = 126$, $R_3 = 243.5$

$$H = \frac{12}{N(N+1)}\left(\frac{R_1^2}{n_1} + \frac{R_2^2}{n_2} + \frac{R_3^2}{n_3}\right) - 3(N+1)$$

$$= \frac{12}{33(33+1)}\left(\frac{(191.5)^2}{11} + \frac{(126)^2}{11} + \frac{(243.5)^2}{11}\right) - 3(33+1)$$

$$\approx 6.741$$

(d) Because $H < 9.210$, fail to reject $H_0$.

(e) There is not enough evidence at the 1% level of significance to conclude that the distribution of ages of the doctorate recipients in at least one field of study is different from the others.

11. (a) The claim is "there is a correlation between the overall score and the price."

$H_0$: $\rho_s = 0$ ; $H_a$: $\rho_s \neq 0$ (claim)

(b) The critical value is 0.829.

(c)

| Score | Rank | Price | Rank | $d$ | $d^2$ |
|---|---|---|---|---|---|
| 93 | 6 | 500 | 5.5 | 0.5 | 0.25 |
| 91 | 5 | 300 | 4 | 1 | 1 |
| 90 | 4 | 500 | 5.5 | −1.5 | 2.25 |
| 87 | 3 | 150 | 2 | 1 | 1 |
| 85 | 2 | 250 | 3 | −1 | 1 |
| 69 | 1 | 130 | 1 | 0 | 0 |
| | | | | | $\sum d^2 = 5.5$ |

$$r_s = 1 - \frac{6\sum d^2}{n(n^2-1)} = 1 - \frac{6(5.5)}{6(6^2-1)} = 0.8429$$

(d) Because $r_s > 0.829$, reject $H_0$.

(e) There is enough evidence at the 10% level of significance to conclude that there is a correlation between the overall score and the price.

13. (a) The claim is "the stops were not random by gender."

    $H_0$: The traffic stops were random by gender.

    $H_a$: The traffic stops were not random by gender. (claim)

    (b) $n_1$ = number of $F$'s = 12

    $n_2$ = number of $M$'s = 13

    lower critical value = 8
    upper critical value = 19

    (c) $G = 14$ runs

    (d) Because $8 < G < 19$, fail to reject $H_0$.

    (e) There is not enough evidence at the 5% level of significance support the claim that the traffic stops were not random by gender.

## CHAPTER 11 QUIZ SOLUTIONS

1. (a) The claim is "the median number of annual volunteer hours is 50."

    $H_0$: median is 50 (claim); $H_a$: median $\neq 50$

    (b) Sign test

    (c) The critical values are $z_0 = \pm 1.96$.

    (d) $x = 23$

    $$z = \frac{(x+0.5) - 0.5n}{\frac{\sqrt{n}}{2}} = \frac{(23+0.5) - 0.5(70)}{\frac{\sqrt{70}}{2}} = -2.75$$

    (e) Because $z < -1.96$, reject $H_0$.

    (f) There is enough evidence at the 5% level of significance to reject the organization's claim that the median number of annual volunteer hours is 50.

2. (a) The claim is "there is a difference in the hourly earnings of union and nonunion workers in state and local governments."

    $H_0$: There is no difference in the hourly earnings.

    $H_a$: There is a difference in the hourly earnings. (claim)

    (b) Wilcoxon rank sum test

    (c) The critical values are $z_0 = \pm 1.645$.

(d)

| Ordered data | Sample | Rank |
|:---:|:---:|:---:|
| 20.45 | N | 1 |
| 21.20 | N | 2 |
| 21.40 | N | 3 |
| 22.05 | N | 4 |
| 22.25 | N | 5 |
| 22.50 | N | 6 |
| 23.10 | N | 7 |
| 24.75 | N | 8 |
| 26.15 | N | 9 |
| 26.75 | U | 10 |
| 26.95 | N | 11 |
| 27.35 | U | 12 |
| 27.60 | U | 13 |
| 27.85 | U | 14 |
| 28.15 | U | 15 |
| 29.05 | U | 16 |
| 29.75 | U | 17 |
| 32.30 | U | 18 |
| 32.88 | U | 19 |
| 35.52 | U | 20 |

$R$ = sum ranks of nonunion workers = 56

$$\mu_R = \frac{n_1(n_1 + n_2 + 1)}{2} = \frac{10(10 + 10 + 1)}{2} = 105$$

$$\sigma_R = \sqrt{\frac{n_1 n_2 (n_1 + n_2 + 1)}{12}} \sqrt{\frac{(10)(10)(10 + 10 + 1)}{12}} \approx 13.229$$

$$z = \frac{R - \mu_R}{\sigma_R} \approx \frac{56 - 105}{13.229} \approx -3.70 \ (or \ 3.70)$$

(e) Because $z < -1.645$, reject $H_0$.

(f) There is enough evidence at the 10% level of significance to support the claim that there is a difference in the hourly earnings of union and nonunion workers in state and local governments.

3. (a) The claim is "the distribution of the sales prices in at least one region is different from the others."

$H_0$: The distribution of the sales prices is the same in all four regions.

$H_a$: The distribution of the sales prices in at least one region is different from the others. (claim)

(b) Kruskal-Wallis test

(c) The critical value is $\chi^2 = 11.345$.

(d)

| Ordered data | Sample | Rank | | Ordered data | Sample | Rank |
|---|---|---|---|---|---|---|
| 101.50 | S | 1 | | 154.20 | MW | 17 |
| 102.60 | S | 2 | | 154.70 | W | 18 |
| 112.70 | S | 3 | | 156.20 | W | 19 |
| 116.30 | S | 4 | | 161.90 | W | 20 |
| 121.20 | S | 5 | | 163.20 | MW | 21 |
| 124.30 | S | 6 | | 166.20 | W | 22 |
| 126.10 | W | 7 | | 167.40 | MW | 23 |
| 127.20 | S | 8 | | 191.90 | W | 24 |
| 132.80 | MW | 9 | | 227.60 | NE | 25 |
| 138.10 | W | 10 | | 228.30 | NE | 26 |
| 138.20 | MW | 11 | | 235.20 | NE | 27 |
| 142.20 | S | 12 | | 239.70 | NE | 28 |
| 142.50 | W | 13 | | 242.20 | NE | 29 |
| 147.60 | MW | 14 | | 249.10 | NE | 30 |
| 149.40 | MW | 15 | | 255.60 | NE | 31 |
| 151.20 | MW | 16 | | 259.90 | NE | 32 |

$R_1 = 228$, $R_2 = 126$, $R_3 = 41$, $R_4 = 133$

$$H = \frac{12}{N(N+1)}\left(\frac{R_1^2}{n_1} + \frac{R_2^2}{n_2} + \frac{R_3^2}{n_3} + \frac{R_4^2}{n_4}\right) - 3(N+1)$$

$$= \frac{12}{32(32+1)}\left(\frac{(228)^2}{8} + \frac{(126)^2}{8} + \frac{(41)^2}{8} + \frac{(133)^2}{8}\right) - 3(32+1)$$

$$\approx 24.906$$

(e) Because $H > 11.345$, reject $H_0$.

(f) There is enough evidence at the 1% level of significance to conclude that the distribution of the sales prices in at least one region is different from the others.

4. (a) The claim is "there is a correlation between the number of emails sent and the number of emails received."

   $H_0: \rho_s = 0$ ; $H_a: \rho_s \neq 0$ (claim)

(b) Spearman rank correlation coefficient.

(c) The critical value is 0.833.

| Emails sent | Rank | Emails received | Rank | $d$ | $d^2$ |
|---|---|---|---|---|---|
| 30 | 8.5 | 32 | 8 | 0.5 | 0.25 |
| 30 | 8.5 | 36 | 9 | -0.5 | 0.25 |
| 25 | 4.5 | 21 | 3 | 1.5 | 2.25 |
| 26 | 6 | 22 | 4.5 | 1.5 | 2.25 |
| 24 | 3 | 20 | 1.5 | 1.5 | 2.25 |
| 18 | 1.5 | 20 | 1.5 | 0 | 0 |
| 18 | 1.5 | 22 | 4.5 | -3 | 9 |
| 25 | 4.5 | 23 | 6.5 | -2 | 4 |
| 28 | 7 | 23 | 6.5 | 0.5 | 0.25 |
| | | | | | $\sum d^2 = 20.5$ |

$$r_s = 1 - \frac{6 \sum d^2}{n(n^2 - 1)} = 1 - \frac{6(20.5)}{9(9^2 - 1)} \approx 0.829$$

(e) Because $r_s < 0.833$, fail to reject $H_0$.

(f) There is not enough evidence at the 1% level of significance to conclude that there is a correlation between the number of emails sent and the number of emails received.

5. (a) The claim is "days with rain are not random."

    $H_0$: The days with rain are random.

    $H_a$: The days with rain are not random. (claim)

(b) Runs test

(c) $n_1$ = number of $N$'s = 15

    $n_2$ = number of $R$'s = 15

    lower critical value = 10
    upper critical value = 22

(d) $G = 16$ runs

(e) Because $10 < G < 22$, fail to reject $H_0$.

(f) There is not enough evidence at the 5% level of significance for the meteorologist to conclude that days with rain are not random.

# Alternative Presentation of the Standard Normal Distribution

## Try It Yourself Solutions

1. (1) 0.4857
   (2) $z = \pm 2.17$

**2a.**

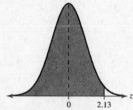

**b.** 0.4834

**c.** Area = 0.5 + 0.4834 = 0.9834

**3a.**

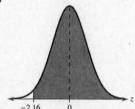

**b.** 0.4846

**c.** Area = 0.50 + 00.4846 = 0.9846

**4a.**

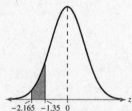

**b.** $z = -1.35$;  Area = 0.4115

**c.** $z = -2.165$;  Area = 0.4848

**d.** Area = 0.4848 − 0.4115 = 0.0733

**e.** 7.33% of the area under the curve falls between $z = -2.165$ and $z = -1.35$.

# Normal Probability Plots and Their Graphs

## Try It Yourself Solutions

**1a.**

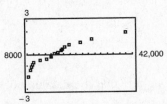

The points do not appear to be approximately linear.

**b.** Because the points do not appear to be approximately linear and there is an outlier, you can conclude that the sample data do not come from a population that has a normal distribution.

## Appendix C EXERCISE SOLUTIONS

1. The observed values are usually plotted along the horizontal axis. The expected $z$-scores are plotted along the vertical axis.

2. If the plotted points in a normal probability plot are approximately linear, then you can conclude that the data come from a normal distribution. If the plotted points are not approximately linear or follow some type of pattern that is not linear, then you can conclude that the data come from a distribution that is not normal. Multiple outliers or clusters of points indicate a distribution that is not normal.

3. Because the points appear to follow a nonlinear pattern, you can conclude that the data do not come from a population that has a normal distribution.

4. Because the points are approximately linear, you can conclude that the data come from a population that has a normal distribution.

5.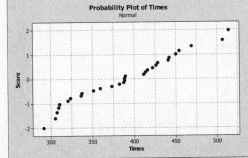

Because the points are approximately linear, you can conclude that the data come from a population that has a normal distribution.

6.

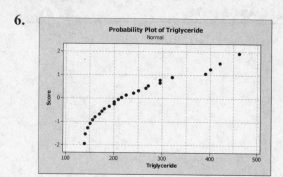

Because the points appear to follow a nonlinear pattern, you can conclude that the data do not come from a population that has a normal distribution.